An Irish Commune

COOPERATIVE STUDIES

BOOKS IN THIS SERIES

1 A.E., *Cooperation and Nationality*
2 A.E., *The National Being*
3 Horace Plunkett, *Ireland in the New Century*
4 E.T. Craig, *An Irish Commune*

EDWARD THOMAS CRAIG

An Irish Commune

THE EXPERIMENT AT RALAHINE, COUNTY CLARE
1831 – 1833

With essays by
James Connolly (1910)
and
Cormac Ó Gráda (1974)
and a foreword by
Terence O'Brien

IRISH ACADEMIC PRESS

This book is published by Irish Academic Press in association with The Society for Cooperative Studies in Ireland. The core of the book is a reprint of *An Irish Commune: The History of Ralahine*, adapted from the narrative of E.T. Craig with an introduction by George Russell and notes by Diarmuid O Cobhthaigh (Dublin, n.d., c.1920). The text by James Connolly is reproduced by kind permission of New Books Publications from their edition of Connolly, *Labour in Irish History* (1973); the publishers also gratefully acknowledge permission given by the Editor, *Irish Economic and Social History*, and by the author, to publish Dr Ó Gráda's article from Volume I (1974).

This edition was printed in the Republic of Ireland for Irish Academic Press, Kill Lane, Blackrock, Co. Dublin.

ISBN 0 7165 2349 3

Our thanks are due to Mr John Donnellan for his kindness in lending a copy of Craig's *An Irish Commune* for the purpose of making this book.

ACKNOWLEDGEMENT

The publishers gratefully acknowledge a grant in aid of publication made by the The Shannon Free Airport Development Company Ltd.

CONTENTS

FOREWORD

In Newtown, on the Welsh borders, beside the River Severn is set a statue of Robert Owen. He stands, arms outstretched, declaiming his views on capitalism, justice and co-operative socialism. These ideas took root in some unlikely places. An Irish landlord, John Scott Vandeleur, applied them to his estate at Ralahine, Co. Clare and persuaded James Edward Craig, who had studied the principles of co-operative colonisation, to come from Manchester and take charge of the project. This book is an edited version of Craig's account of life on the commune from the time he arrived in the Spring of 1831 to the abrupt and calamitous dispersal of the community some two and a half years later.

On 23 November 1833 the members of the Ralahine Agricultural and Manufacturing Co-operative Association met for the last time and placed on record a declaration of 'the contentment, peace and happiness they had experienced . . . under the arrangements introduced by Mr Vandeleur and Mr Craig which through no fault of the Association was now at an end.'

So ended an extraordinary social experiment, a drama so poignant with quiet heroism and tragedy that it resembles a nineteenth century Russian novel. It is not at all surprising that Margaretta D'Arcy and John Arden adapted it for the stage as *Vandeleur's Folly* in 1978.

Craig's book is one of a series of reissues undertaken by the Society for Co-operative Studies in Ireland. These are intended to provide students of co-operation with works of historical interest which have become generally unavailable. Because Ralahine was a source of inspiration to two great Irish social thinkers in this century, George Russell (AE) and James Connolly, it was considered useful to include some of their thoughts on the subject. Consequently, the preface by AE to the edited reissue of Craig's book in 1920 has been retained, and chapter eleven from Connolly's *Labour in Irish History* (1910) has been added as an appendix.

If Vandeleur's behaviour had not led to the dissolution of the commune could it have survived long enough to provide an enlightened model for other landlords on the island? This is a question that has often been raised. The editor of the Journal of the Economic and Social History Society of Ireland has kindly permitted the inclusion of an article by Dr Cormac Ó Gráda, as an appendix, which analyses the venture from the perspective of an economic historian and concludes that survival was in some doubt, given the favourable agricultural conditions necessary to sustain the high rental.

What relevance then has the history of Ralahine to the contemporary co-operative movement in general and to rural Ireland in particular?

Firstly, it establishes formal agricultural co-operative organisation in Ireland fifty years before Plunkett came on the scene. It is also a striking example of what could be achieved by the vigorous application of co-operative principles of organisation. Illiterate, poverty-stricken, murderous labourers were transformed into a purposeful, hardworking and self-sustaining community. The circumstances in which these principles were successfully applied were peculiar to that particular situation and time. However, the same philosophy has in the last twenty years given rise to many community development co-operatives in rural areas, in towns and latterly in the cities. They are to be found inside and outside the Gaeltacht and north and south of the border. In a country with no official administrative structures at parish level they often constitute a type of surrogate local government and initiate collective approaches to a wide range of investment and social provision. Individual agencies of government have recognised their potential and are encouraging their establishment.

The co-operative scheme at Ralahine was imported from without and inspired from above. While informal co-operative groups were commonly formed in the Irish countryside to accomplish specific tasks, co-operation as a philosophy and a working body of principles was new. The success which the venture enjoyed was mainly due to the character and personality of the manager, Edward Thomas Craig. Today, formal co-operative organisation is an established part of Irish life, but the success or failure of the present generation of community co-ops still rests, as at Ralahine, largely on the calibre of their respective managers.

It has been claimed on behalf of some types of co-operatives that they have provided a receptive environment to the introduction of new technology. Bolger in his history draws attention to the fact that the first mowing machine in Ireland was in operation at Ralahine. It is interesting to note that, in our own day, Corca Dhuibhne Community Development Co-op in Kerry was one of the first enterprises to reclaim land by using a special deep plough to crack the hard pan, with beneficial economic consequences for the smallholders in the area.

The Ralahine experiment does not seem to have had any appreciable influence on the Ireland of its day. In our own time the name has come to have a symbolic and sentimental significance; a reminder that co-operative organisation can bring a benign influence to people's lives. This is illustrated by the Ralahine-named estate built by a housing co-operative in Ballybrack, Co. Dublin. That the experiment still influences Irish thought is instanced by the arguments recently put forward by David Lee who sees the co-operative as an attempt to disguise class conflict, which he links with the contemporary issue of worker-participation.

> Although it is 150 years since the Ralahine Co-operative was founded the lessons of this Irish social experiment are relevant today. Workers should beware of worker participation schemes that are handed down to them by employers for in the long term such schemes will injure the interests of the trade union movement and themselves.[1]

One of the most affecting parts of the book is the description of the distress of the workers and their families on

1. *Ralahine Land War and the Cooperative*, Limerick: The Bottom Dog; Dublin: Co-op Books; 1981, p.56.

learning of the collapse of their hopes. Their pain and despair carries over the years and touches the reader deeply. Rural Ireland in the eighteen thirties was a society in crisis, struggling to adjust to new economic forces which were making agricultural labourers redundant. The experiment at Ralahine was a response to that situation. The members of the commune had good cause to lament its passing for by changing the economic and social framework within which these labourers functioned, the co-operative had been able to offer them continuous employment and a measure of social security at a time when rural underemployment was endemic.

Contemporary Ireland is also going through an adjustment process in response to severe economic difficulties and over a quarter of a million people on the island are now registered as being without work. Many of these are without financial or educational resources. In economic terms they are marginal people, the urban cotters of the late twentieth century subsisting on the new potato, social welfare. Edward Craig's philosophy and his achievement are relevant today in so far as they challenge us to reconsider the capacity of communities to create local employment opportunities in socially useful activities by means of co-operative forms of organisation that alter the economic and social framework within which such work is offered. Co-operative philosophy reminds us that between market capitalism and state socialism there is a third way.

The Society wishes to thank The Shannon Free Airport Development Company Ltd for the generous sponsorship which enabled this re-issue to appear, and Irish Academic Press, which has undertaken to publish this series.

TERENCE O'BRIEN, Chairman
SOCIETY FOR CO-OPERATIVE STUDIES IN IRELAND
23 November 1982

INTRODUCTION.

We cannot say of Ralahine that the seed fell on stony ground, but we might almost say that thieves did break through and steal. No Irishman can read the story of this famous experiment without the most poignant regrets that it was not allowed to continue. When John Scott Vandeleur gambled at his club he gambled away not merely his own property but what may well have been a happier destiny for his country. It is inconceivable that if the community founded at Ralahine had developed as it began it would not have affected the rest of Ireland. It might have saved us many years of tragic history, and instead of beginning our agricultural co-operation long after Denmark, Germany, and France we might have been the pioneer nation. Nobody can read this narrative without feeling affection for the historian. A true, kind, stout-hearted man this Craig, with an infinite belief in the magic of goodwill and justice, with a wisdom and tolerance he would have us believe arose from adherence to advanced philosophical principles, but which we feel came from a heart naturally sympathetic. It needed bravery to emigrate to Clare in 1830. We see no reason to doubt that Craig's account of political and social conditions was other than restrained. He came to that savage and turbulent society as confident, we may well believe, in the essential goodness of human nature as in his economic theory. He seems to have been a good business man and from the economic point of view the co-operative plan he devised was successful.

Visitors to this strange community in Clare found that the members had a higher standard of comfort than the highly-paid operatives in Lancashire, yet all about Ralahine were rags and poverty, murders and evictions, political fury and coercion. We do not rely on Craig alone for knowledge of the state of Ireland at the time, nor on him alone for knowledge of the changes brought about at Ralahine. As I read Craig's narrative my thoughts ran on another instance of the magic of sympathy. One bitter winter in Moscow, Tolstoi tells us, a committee had been formed to investigate conditions among the very poor. In pursuance of his duty as a member he went into a brothel guided by a police officer. No one there took any notice of his coming. All the women were apathetic in their wretchedness. The woman Tolstoi spoke to made no answer. Then the police officer interrupted with a harsh command to this miserable creature to speak. Tolstoi reproved him. "It is we," he said, "who have made these women what they are." The accent of pity never before heard there had a magical effect. All around the room rose up touselled heads and staring faces. That sudden awakening, said Tolstoi, reminded him of nothing so much as the vision of the valley of bones in Ezekiel reanimated with life at the breath of the Spirit. Not so swiftly but as surely by this power of sympathy the wise and tolerant Craig wrought his miracle at Ralahine. It is a power which rulers who seek for power most often ignore, though by it empires and even autocracies might be built enduringly. While Craig was winning men from violence by reason and justice, the rulers of Ireland were then as ever relying on physical force, accepting policies from policemen who are the power behind the throne in Ireland, the superstatesmen to whom Chief Secretaries in trouble infallibly turn, accepting their guidance even when economic problems relating to land and labour, or spiritual problems arising out of the Gaelic and national culture of Ireland are considered. Ireland in Craig's day was asking for bread and it was given a baton. We notice that no Chief Secretary came to study what was being done at Ralahine, though a word from one of the powerful ones in praise of the co-operative community might have turned the thoughts of hundreds of landowners to a solution of the agrarian problem which promised not only peace but plenty.

Conditions have changed in Ireland. It would perhaps be impossible to use the rules devised by Craig in a co-operative community to-day. But we are moving rapidly to the creation of co-operative communities all over Ireland. We have co-operation in purchase, manufacture and sale in every county, and I hear with pleasure that societies for co-operative farming on the Italian model will soon be established, and these Italian co-operative land cultivation societies are the nearest things in Europe to the Ralahine community; and in Italy I am told they regard that long-vanished society at Ralahine as the pioneer or morning star of their own fascinating movement. We have passed beyond some of Craig's ideas, but we have not yet attained to his central idea of a community feeding itself, clothing itself, educating its children, applying its surpluses to further development. There was something of the imaginative seer and poet in Craig. He asks a question which many are considering, and which sometime the workers of the world will ask and insist on a true answer. "Why should not the builders of palaces live in them? Why cannot the joiner, the mason and other artisans dwell in mansions? The answer is given by Ralahine. It is simply a matter of association, labour, culture, time and capital." Yes, surely the spirit in man will ask for these things in due time and win them for itself and much more, for the soul of man is immortal and exceeding high, so high that, if by a fragile thought, a coloured dream, a thing of air its immortality is captured, it must fare on that adventure, holding to its purpose through centuries until it has wrought in massy gold what was at first but an evanescent breath.

With such ideas in his mind we can understand the bitterness in the heart of Craig when he learned of the flight of John Scott Vandeleur, and the destruction of his dream. "My own feelings were much disturbed. The heart was too full. I felt as weak as a child. A gush of tears gave relief." It was deep and enduring this feeling. Years afterwards when he was writing the history of Ralahine he says, "Feelings are awakened by the echoes of the past which now fall upon the ear and cause tears to flow and wet the paper on which I write." Perhaps he had spoken of these high dreams of his to the members of the community and to that flighty idealist, Vandeleur, who may have encouraged the community to hope for their realisation, for is there not something of the agony of a defeated dream in the keening of the community hearing of the vanishing of their chief? "Ohone! Ohone! Shawn Vandeleur, why did you go from us? Ohone, Vandeleur! why did you leave us? Why have you left your own Ralahine?" We can imagine that flighty idealist now listening to Owens or Craig, moved sincerely by them, and then as excited over the dice or cards and, losing all, going out into the darkness, perhaps thinking most bitterly at the end not of the loss of land or money but of the dream his passions prevented from becoming true.

I think it possible that the soul of Ralahine may reincarnate speedily in modern Ireland where people now talk almost as much about co-operation as about Sinn Fein. They are better prepared for communal activity. The republication of Craig's book will help to crystallise ideas which at present are in solution, and I consider it the best of all good omens that this week before I began to write I heard that another landowner in Clare has just begun another communal experiment on his property. It is right that the idea should be born again in Clare and I hope there will arise a more fortunate Ralahine at Raheen. Æ.

CHAPTER I.

IRELAND AND THE IRISH

The population of Ireland in the year 1830 amounted to about seven-and-a-half millions. Poverty is marvellously prolific, even when want grows faster than food. Land in Ireland available for tillage is limited, and as it forms the basis of existence, it becomes an object of great vital importance to obtain it, and hence competition had raised its value while it reduced the wages of labour and the means of subsistence. The food of the peasantry consisted chiefly of potatoes. In the south and west the crop had failed. To add to the evil, large landlords had begun to evict their tenants, and to reduce small holdings, while, owing to the want of capital and of confidence, tillage lands were converted into grazing farms, on which a herdsman and boy could supersede some twenty labourers and ploughmen. The rents of con-acre were both enormous and unfair, from the fact that the poor tenant supplied the manure for the potato crop, while the landlords took the benefit in the grain crops subsequently sown. Rents were demanded at £8, £10, and in some cases at £14 per acre. If the tenant could raise a sufficient crop of potatoes to pay the rent and sustain his family he considered himself fortunate. In many cases the crops were taken to the market attended by the agents, and the proceeds handed to them in payment of rent, while the slave of toil returned home empty-handed, with the galling knowledge that the fruits of his labour were taken by another, who was perhaps the representative of an absentee. In bad seasons famine soon became prevalent. The labourer and his family under such conditions were doomed to want and starvation. Peace and order under such circumstances were impossible. Coercion Bills, Arms Bills, an armed police force of 30,000 men, and a large proportion of the British army, might make a solitude, but that would not make peace, order, and contentment.

Under the conditions indicated many perished in silence, while thousands, alike ignorant of the causes of their misery and of the remedy, banded together in the vain hope of finding a cure for their sorrows by striking terror into the great landlords, their agents, and the Government. The distrust between the landholder, the tenantry, and the labourers, soon became manifest and alarming, and so dangerous towards the end of that year that in December, the Lord Lieutenant (the Marquis of Anglesey) was urgently solicited by the magistrates to visit County Clare to aid in the suppression of armed bands, perpetrating the most violent acts of lawlessness. This was, in fact, an acknowledgment that the local authorities, although aided by an armed police force, were unable to preserve the peace or protect the lives of the wealthy landlords.

The starving peasantry were clamorous for land, for employment, and for food, and it was proposed to convince by military force of argument, and to silence them by a supply of powder and shot. While quietness followed the appearance of the soldiers and the police in one district, the discontented and starving peasantry assembled together in other parts and perpetrated many atrocious crimes.

An active magistrate, an obnoxious landholder, agent,

or steward, was not assured of safety for a single night The peasantry marched in bands through the south-western counties, demanding a reduction of rents and an increase of wages, which were then only sixpence per day for agricultural labourers.

In some districts they insisted that spade husbandry should be the mode of tillage. In other places they took labourers from their work and horses from the ploughs. Arms were obtained by breaking into houses during the night, and the weapons which everyone deemed necessary for his own safety were carried away.

Many of the landlords fled in terror, as was natural for them to do, and left their mansions in the care of the police. Destitution, want, distress, and misery were widely spread, and no practical help at hand. In the west of Ireland there were 200,000 persons in want of food and the means of obtaining it. Parliament was applied to for relief and the Chancellor of the Exchequer proposed a vote of £50,000 to be applied to the starving poor in making roads. In the meantime there were many in whose hearts Famine had written fiend.

Agrarian outrages, murder, robbery, and searching for arms were perpetrated in groups, which could not be controlled by the police. Although the police were supplied with firearms, and were under military discipline, they were unable to cope with the discontent and violence manifested at various points. At Carricshaugh, the police attacked a party when proceeding to enter a house in search of arms, when five of the police were shot dead, and several of the rioters killed and wounded.

The next movement of the peasantry was agains pasture land, as it deprived them of employment. No tillage, no labour. Men and women armed themselves with their agricultural implements, and, in open day, in savage desperation, dug up the grass lands, levelled the wall fences, and turned the cattle adrift. In many instances the animals were houghed, and so maimed that they had to be killed and sold. On one occasion a number of peasants had assembled on a grazing farm, levelling the boundary walls, and turning up the grass, when a troop of soldiers, accompanied by one or two magistrates, made their appearance. The peasantry stood their ground; the Riot Act was read. They said they might as well be slaughtered as starved to death from the want of work. It was a painful spectacle. It was for bullocks rather than brains; the landless labourer against the landlords and land-grabbers. A terrible tragedy seemed inevitable, when a parley was advised, and one of the magistrates appealed to the men to desist and adopt other modes of seeking redress for their wrongs. The peasantry still continued to meet in the night, began to levy blackmail on the small farmers, and had complete control over large portions of County Clare. This condition of the people was subsequently admitted in Parliament during the discussion on the Coercion Bill of Earl Grey's Government.

The Catholic priests formed a committee to try and restore peace, order, and tranquillity; but they soon found themselves powerless to subdue the outbursts of discontent, and made known by public announcement their utter inability to arrest the prevailing system of outrage and murder.

A popular advocate of Repeal of the Union, a friend of Daniel O'Connell—Mr. Thomas Steele—put forth an address, which was posted up in Ennis and other places, in which he declared against the White Boys and Terry Alts, thus: "Unless you desist, I denounce you as traitors to the cause of liberty in Ireland. . . . . I leave you to the Government and the fire and bayonets of the military. Your blood be upon your own souls." This denunciation did not meet the demand for work or supply wages or food, and the famishing labourers continued to throw a lurid light on the social chaos around them.

At the request of the magistrates, the Lord Lieutenant made a progress through County Clare in May, accompanied by a display of military force and artillery, and attended by the magistrates. A piece of ordnance was planted on the bridge of Clare, ready to exterminate the insurgent peasants. His lordship was received with passive indifference.

These military displays have a ghastly aspect when viewed as social remedies. The labourers ask for work and they are offered cold steel and lead seasoned with "villainous saltpetre." The evil days still continued; want was thinning the bodies and the ranks of the poor. Proclamation of the "Insurrection Act" was duly made; and although all were enjoined to keep within their houses after dark, midnight assemblages were still continued. A Special Commission tried and condemned several miserable wretches for murder, carrying arms, digging up ground, and other violent outrages. But even while the Commission was sitting night visits were made by "Terry Alts," "Whitefeet," and "Lady Clare's Boys."

CHAPTER II.

FEAR SEIZED UPON ALL HEARTS

From the brief statement given of the condition of the people in County Clare, in 1831, it is very evident the law was powerless for the preservation of life and property. Want and destitution had made the peasantry desperate, till they began to despise the police force, and defied the Government, whose military display had little terror for starving men, while the Catholic priesthood had no moral control over discontented people made furious by famine.

Here was a melancholy picture of a rich soil only partially cultivated, and a willing people unemployed. Condemned to remain ignorant, they had become brutal in their revenge for social injustice, and driven to wild and demoniacal deeds of desperate violence through lack of food, work, and useful employment. Goaded to resistance by centuries of wrong and bad Government, they were aroused to fury by famine, and now showed "the dragon's teeth" with a vengeance. The only measures that were popular as remedies were military and police "repression," legal penalties, and religious influence. To preserve property it was sagely deemed necessary to shoot and destroy the creators of wealth. Fear and suspicion seized upon all hearts, all classes seemed paralysed and utterly powerless and incapable of suggesting, let alone adopting, a remedy.

It was at this time the spirit of vengeance reached Ralahine, and the family of the proprietor were obliged to leave their mansion in the charge of an armed police force, and seek safety in the city of Limerick. This circumstance prompted the proprietor, John Scott Vandeleur, Esq., then late High Sheriff of the County Clare, to carry into action a desire he had long cherished of establishing a co-operative farm on his property at Ralahine. The condition of the peasantry urged him to make the change sooner than he had intended.

Although Mr. Vandeleur had resolved to make a change, the difficulty was how to carry into practice the contemplated arrangement. He could not effect it himself, although he saw clearly what he would like to realise. He knew no one in Ireland who could or would help him. He came to England to seek for assistance, and was advised to apply to me to organise the peasantry into an agricultural and manufacturing association, to whom he would be willing to let the estate of Ralahine as a co-operative farm.

Looking at the condition of the people then, as reported in the South of Ireland, and at the number of murders and agrarian outrages in County Clare, I was doubtful of success among a people in a state of insurrection, and requested a little time for consideration before I finally decided to accept the mission.

According to the opinions of my friends, there could be little prospect of success with the White Boys in their then state of insubordination and outrageous violence. My family was strongly opposed to the undertaking, and the possible sacrifice of future interest in a legacy was intimated. I was not swayed by any consideration as to the loss of the prospective legacy, nor did I appreciate the objections urged on account of the plan being impracticable, Utopian, and absurd, because it implied a rash, perverse, and obstinate spirit, instead of a calm and deliberate investigation of the conditions necessary to success. I fully appreciated the difficulties, and I had confidence that, with prudence and perseverance, these might be overcome and conquered.

I arrived at the conclusion that an association to rent a farm, stock, and buildings was practicable and desirable; and that if suitable productive efforts were sustained by intelligent guidance, and agreement to give the labourer a participation in the net profits over rent and interest of capital employed, the system would realise success for all concerned. It was anticipated also, that if successful in Ireland, the example would exercise some influence over the movement, as it is now doing in England and America.

On calling on Mr. Vandeleur, who was waiting for my decision at his hotel then in existence at the corner of King-street, Manchester, and finding him prepared to carry out certain conditions which I proposed, and being strongly impressed with the frankness and apparent sincerity of his character, I consented to leave England and go to Ireland as soon as necessary. He thought it desirable to be at Ralahine as soon as possible to prepare the people and frame the regulations and make the organisation necessary for the contemplated change.

On arriving at Limerick, I had the first glimpse of the noble river Shannon, which takes its rise in Lough Derg, and passes through the heart of the island.

The neglected state of the splendid river Shannon was typical of the condition of the land and the people of Ireland, and a disgrace to those who claimed to be the "governing classes." This noble river of waters washes the shores of not fewer than ten counties. For sixty miles below Limerick it is the noblest river in the

British Isles, being from one to eight miles broad, affording water carriage from its source to its mouth, and is about five miles wide as it washes the shores contiguous to Ralahine. The absence of cranes on the tidal part of the river caused great impediments to trade and traffic. Ships laden with merchandise would ride as near in-shore as practicable during high tide, and then drop anchor. At low water, porters would take their horses and carts to the vessel, and for a short interval help to discharge the cargo. The tide would soon return, and thus many days would be wasted in waiting for the fall of the tide to resume the work! Hundreds of thousands of pounds had been granted as loans by Government to the landlord class for the improvement of roads, and to carry on public works. Large estates had been improved, but public works had to wait, and the Shannon was then waiting for "Government Help."

CHAPTER III.

SOME OF THE LOCAL SURROUNDINGS OF RALAHINE

On my arrival at Ralahine I found the estate admirably adapted for the purposes of a co-operative farm. It consisted of 618 acres, about one-half of which was under tillage, with suitable farm buildings, and situated between the two main roads from Limerick to Ennis. A bog of sixty-three acres supplied fuel. A lake on the borders of the estate gave a constant and available supply of water power, and a small stream flowing from it gave eight-horse power to a thrashing mill, scutch and saw mill, a lathe, &c. A fall of twenty-horse power was available at a short distance, when required, for manufacturing purposes.

A large building had been erected, 30ft by 15ft., suitable for a dining-hall, with a room of the same size above, suitable for lectures, reading-room, or classes. Close to these were a store-room and a dormitory above. A few yards from and at right angles to the large rooms were six cottages in course of erection. At several hundred yards' distance stood the old Castle of Ralahine, with its lofty square tower and arched floors, capable of being temporarily adapted for the accommodation of those whom it was intended to unite in the proposed association. The estate is about twelve miles from Limerick and about the same distance from Ennis. Newmarket-on-Fergus is three miles distant on one side, and Bunratty Castle the same distance in the other direction.

Although much gratified with the capabilities of the

land and the position and surroundings of the estate, the condition and prejudices of the people in the neighbourhood were not encouraging, and soon became a source of anxiety as to the possibility of success in organising a system of mutual co-operation among them.

The labourers were employed on the usual conditions, and while some few lived outside of the estate, others had to walk a mile; several had to go five or six miles to their homes—a great cause of loss of physical power, comfort, and convenience. They had until lately been managed under the direction of a steward, whom I now heard for the first time had been murdered a short time before my arrival. Mr. Vandeleur had not given this information from a desire not to discourage me. The steward had been somewhat despotic, harsh, and severe in his treatment of the labourers. One scorching hot day the men were at work reaping, and occasionally stopped to take a draught of water to cool their parched thirst, when the steward kicked it over, declaring he would not have the can there to cause the men to lose their time. The steward merely superintended the industrious workers, and had no burning thirst, nor any sympathy with those who suffered under exhausting conditions. Similar acts of unfeeling harshness roused a spirit of vengeance. A midnight meeting was held in Cratloe Wood, on the borders of the lake on the estate, and it was decreed to put the steward to death, and lots were drawn who was to perpetrate the foul deed.

In their savage revenge he was shot dead one night as he was bolting his door, which was opposite to the window where the assassin stood waitting for an opportune moment, when the shot was fired, and two bullets were sent through his skull and marked their mission on the door. The deed was done in the presence of his wife, to whom he had only been married three months. It was known on that fatal morning that vengeance was to be executed, and the report of the gun was heard by several who were at the farmstead, which was not far distant, but the assassin was allowed to escape and was never brought to justice. This event created great alarm, and, as already stated, the proprietor's family, in fear, left their mansion in the care of an armed police force. As a result of the murder the peasantry became divided into the party of the late steward's family and friends, and those of the Terry Alts or labourers.

Arriving at this juncture, I was received with a good deal of suspicion (being a stranger and a *Sassenach*) by the people in the neighbourhood, for they were not aware of the exact purpose in view. Being the only Saxon in that part of Ireland, and arriving while the people were in a state of wild frenzy of indignation against their forced exclusion from the soil, they naturally concluded that, as traditionally all Sassenachs were incapable of dealing fairly and justly towards Irishmen, I should secretly sympathise with the landlords and the police authorities. Their prejudices and suspicions led them to suspect me as likely to betray them by obtaining the name of the man who had murdered the steward. One or two of the more intelligent artisans on the estate had more confidence in my good intentions, and while they were forced to take the secret oath of resistance to the authorities, I was forewarned by them of events which were likely to frustrate me.

Although Daniel O'Connell and other politicians were denouncing the Government and advocating a Repeal of the Union as the only remedy for the wrongs and sufferings of Irishmen, I soon discovered reasons for ascribing the disorganised condition of the people to agrarian injustice, social degradation, and suffering and discontent brought on by the legislation of the past, as the Statute Book was filled with scorpions in the form of vindictive penalties and Coercion Bills. As Arthur Young declared, in 1776, "the Legislature passed Acts of Parliament which became the meridian of Barbary." The results were now manifest. The armed peasantry resisted the military force, prevented the roads being repaired, and several policemen were shot while attempting to resist the proceedings of the insurgents and the starving labourers.

The whole of the County of Clare had been declared, on the 10th of May, 1830, to be in a state of disturbance requiring extraordinary establishments of police, and was placed under the Peace Preservation Act. But these were mere measures of repression, and did not provide any remedy for the distress, disturbance, and insurrection. The wretched and half-famished peasantry assembled in multitudes, and with the rudest kind of weapons of warfare demanded liberty to labour and permission to till the lands, on which they were born, for subsistence.

The landlords urged the Lord Lieutenant to disarm the people. Lord Anglesey said he would not call upon the people of Clare to give up their arms until the gentry did something for their relief. Under these conditions the leading landlords drew up the following series of inquiries, to be submitted to the landlords and gentry, requesting them to offer suggestions to be laid before the Lord Lieutenant. One of these circulars was submitted to me.

TO JOHN SCOTT VANDELEUR, ESQ.

Sir,—The pledge given to His Excellency the Lord Lieutenant (the Marquis of Anglesey), when in Clare, that as soon as good order was restored the gentlemen of the County would take into consideration the situation of the labouring classes of the community, induces a request that you would commit to writing your opinions upon that subject, with a view of recommending for general adoption any measures which you might think conducive to the greater diffusion of comfort among the people.

It is not intended to prescribe to you, in the slightest degree, as to the points to be considered, or the order of treating them; but the following are respectfully submitted, as coming within the range of the proposed investigation, which should comprehend the duties and interests of head landlords, middlemen, land agents, tenants, and labourers, who, having no land are obliged to resort for their subsistence to con-acres.

1. Leases, whether they should be given; their duration; covenants; possibility of making the amount of rent dependent on prices during war or peace respectively.

2. Sub-division of land; extent to which it should be carried or permitted; and the conduct which policy and humanity prescribe towards persons brought to farms without the landlord's consent.

3. Profitable employment of a portion of annual rent in improvements of the soil, whereby labour would be afforded to the indigent poor; expenditure of capital in building; increase of rent which the landlord should expect as a remuneration for outlay.

4. Rate of wages in summer and winter, cottage and gardens for the poor, con-acre rent.

5. Equitable claims upon non-resident landlords who afford no employment to the various classes of society dependent on the affluent.

6. Advantages that would result from resident agents of intelligence upon all estates producing a rental of £400 a year, or upwards, in countenancing improvements, enforcing covenants in leases, and other ways.

There are, doubtless, other points to be considered, which long experience of the county will suggest, and any observations you may be pleased to favour me with may be sealed up in a packet endorsed "Observations for the Consideration of the Landed Proprietors of Clare," and left with Mr. Knox, in Ennis, on or before the 1st of November. Your name may be attached or not, at your own discretion.

An endeavour will be made to form a digest of the observations of the different gentlemen who have been requested to contribute, and publicity will be given to them in the manner which may hereafter appear expedient.

I have the honour to be, &c.,

JAMES MOLONY.

Sept. 21, 1831.

This circular clearly indicates the fact that the great landlords of County Clare admitted the existence of the causes which had resulted in widespread discontent and rebellion against the sufferings and wrongs which the landlords themselves inflicted and could easily have removed. They indicate the propriety of claiming an increased rent for cottages built with their capital, but omit all reference to the fact that when a farmer is evicted the landlord claims all buildings and improvements realised by the tenant's labour as his property, and raises the rental to the succeeding tenant—a robbery and an injustice allowed by the Legislature.

This document also establishes the fact that neither the Government, the landlords of Ireland, nor the wealthy classes understood or admitted the real cause of the evils existing around them, nor did they comprehend the remedies required. As Lord Bacon wisely said, "to allay sedition we must allay the maker of it." Political economists advised the reduction of small holdings, increase of large farms, and the removal of the surplus population by the emigration of the people.

The conclusions to which the relations of the labourer to the land and the fruits of his toil led me were that the causes at work were social and agrarian, as well as political; and that social amelioration and a share in the net profits, if any, after paying rent and the interest of capital, would realise a great change at once in the spirit and ameliorate the condition of the people.

There are in Ireland 1,500 absentee landlords, possessing 3,200,000 acres, who live in England, London, Paris, or other fashionable rendezvous. There are also 4,500 landlords, owning 4,200,000 acres, who do not reside on their estates, but in Dublin. The rents abstracted from these tenancies are a tax upon toil. We ought to be more surprised at the submission of the people than at their indignation and discontent.

The state of the people and their demand for relief at that time in County Clare is forcibly illustrated by the statements made at the opening of the Reformed Parliament. The King in his speech pointed to the fact that Ireland was in a state of rebellion, and that repression was the only remedy available to Government. Earl Grey, in the House of Lords, moved that the Clerk of the House read that part of the King's Speech relating to Ireland, and in which he declares that "a spirit of insubordination and violence has risen to the most fearful height, rendering life and property insecure, defying the authority of the law, and threatening the most fatal consequences if not promptly and effectually repressed."

The measures of repression proposed were the suspension of the Habeas Corpus Act and the introduction of martial law in the Coercion Bill; Earl Grey said that "The situation in Ireland had become, and now was, worse, perhaps, than at any former period. Bodies of men were collected and arrayed by signals, by ringing of chapel bells, by blowing of horns, by lighting of fires, evidently directed by a system of organisation, and operations in which many were connected, and conducted in a manner, up to the present time, that had defied all the powers of the Government and the law effectually to repress. The disturbers prescribed the terms on which the land should be let, and any who disobeyed their orders were subject to have their property destroyed, and even to the penalty of death. They dictated what persons should be employed, and by whom, forbidding labourers to work for obnoxious masters, and preventing masters from employing such as were not obedient to their orders. They enforced their commands by acts of cruelty and outrage, by spoliation, by murder, by attacks on houses in the dead of the night, by dragging the inmates out of bed, by beating them sometimes to such an extent that death ensued, by inflicting evils that were often scarcely less than death. He found, from January, 1832, to December, 1832, a most frightful catalogue of crimes perpetrated in Ireland. Homicides, 172; robberies, 465; illegal notices, 2,095;

illegal meetings, 425 ; houghing of cattle, 455 ; malicious injuries to property, 769 ; attacks on houses, 763 ; firing at houses, 280 ; serious assaults, 3,156."

Sir Harry Vivian reported, on the 16th September, 1832, that since he had assumed the command of the army, on the 1st of July, 1831, he had received from officers commanding detachments in different parts reports of not less than fifty murders (not including the eleven policemen at Carrickshaugh) that had been committed in the immediate vicinity of their cantonments.

The police had little or no control over these outrages. The number of rewards offered for the discovery of atrocious crimes were 108, and of this number no more than two were paid! This immunity from detection arose from the secret associations and sympathy of nearly the whole of the peasantry with the movement against their exclusion from the land. As Judge Blackburn said, "Land is a necessity of life ; the alternative of not getting it is—starvation."

So well were the proceedings of the White Boys arranged, that almost at the very moment at which the police were ordered out beacons were lighted as signal fires upon the hills to give warnings to the bands out on their nightly mission of destruction and terror.

CHAPTER IV.

TERROR CAUSED BY WHITE BOY DISGUISES

The peasantry formed themselves into secret societies, taking various names, such as "White Boys" and "Lady Clare Boys." The men were sworn upon oath to preserve the secrets of their association, and to act for the interest of their class. The "Whitefeet," or "White Boys" were disguised by wearing a white shirt over their clothes, and a mask. The "Lady Clare Boys" were disguised by wearing women's dresses and masks made of the sides of a black hat, with quills sewed over the hole for the mouth in place of teeth. These disguises were horribly grotesque; and when it was known that firearms were carried under the disguise, the stoutest heart might quail on the approach of these hideous and savage instruments of discontent. In their wild rage they perpetrated acts the bare recital of which made the heart palpitate, One of these men in disguise visited Ralahine to administer their oath to one of the men, who happened to be absent at the time. A carpenter employed in preparing materials for the new houses fled, and sought safety inside the water-wheel of the thrashing mill, which, if the sluice had been opened, would have carried him continuously round and killed him. In one district bands of men, disguised with white handkerchiefs round their waists, were going about swearing the people to pay neither tithes, taxes, nor rents, and beating those who refused.

The Government had no clear perception of any practical remedy beyond the suspension of the Habeas Corpus Act, the temporary abolition of trial by jury, the introduction of a military tribunal, and martial law, as a substitute for the usual courts of justice. O'Connell had great influence over the people of Munster; but he had no remedy beyond "repeal," and denounced the "brutal and bloody Whigs," the promoters of the Coercion Bill, as "men with heads of lead, hearts of stone, and fangs of iron." Ireland was in a condition from which neither the heads of the Church, the army, nor the Government could relieve her. She seemed as if sinking into a Slough of Despond, and waiting for some social Giant Great-heart to raise her from her wretchedness and sorrow, to peace, comfort, and happiness.

My arrival at Ralahine at this time, when secret associations, the "White Boys," "Terry Alts" (labourers), and others, appeared to have full control over the people in defiance of the authorities, was a great disadvantage to me. My attire was in marked contrast with that of their home-spun frieze coats; my language was to them a foreign tongue, and proved me to be a Sassenach; and their traditionary histories of the Saxon are but dark memories of conquest, confiscations, ejectments, injustice, and tyranny. Englishmen who have not resided among the peasantry of Ireland can form no adequate conception of the hereditary bitterness and sense of wrong left by the cruelties, injustice, and sad memories arising out of the statutes of Kilkenny, the "black spot" in the Treaty of Limerick, the "bloody slaughterings at Drogheda," and other well-remembered localities.

These views were more or less prevalent in the South, and had induced some of the peasants to sympathise with and yield to the influence of the secret societies. It was soon very clear that not one of the persons employed on the Ralahine Estate was in favour of any change in dealing with the wage conditions existing around them. The labourers were moody, dubious, and discontented. Owing to suspicion, I was, on one occasion, advised when detained beyond sunset to return to my lodgings by a different route. On another occasion, I was struck with a stone from behind,and on yet another occasion, a rudely-drawn coffin lay in my way.

Although the proclamation of the Lord Lieutenant was issued against illegal assemblages and midnight meetings, yet great numbers of the labourers met and passed along the high roads in defiance of the military and the armed police. On one occasion,at midnight, they met near the cottage of the widow where I lodged and kept up for a time a wild uproar and noise, such as an Irish crowd only can utter. It seemed as if a host of fiends were let loose with a view of committing some desperate deed of death and destruction. The shrill clamour was neither like a wail for the dead nor a cheer for the hope of the living. I had little fear of death, but these fierce outbursts of feeling often imparted a strange emotion and a vague apprehension of violence. I was informed on the day following that the outlines of a grave had been cut by the removal of the grass surface as a warning for someone, but could never ascertain for whom it was intended.

These strange proceedings made my position very irksome, requiring great firmness, prudence, and decision. To add to my difficulties, the proprietor's family were opposed to the contemplated arrangements. The servants in the hall knew that the " new system " was not desired by the family or the gentry, and their vulgar jests and coarse humour had to be tolerated with the best grace possible. I had, however, made up my mind to adapt myself to the circumstances of the people, and endure every inconvenience and all kinds of opposition, so long as I could rely on the proprietor ; but I remember that, one day, while engaged in drawing out the plan and regulations, I heard a horrible account of a murder of a steward in open day, and in presence of the labourers in a field in the neighbourhood, when my feelings were powerfully affected, mastered my constrained emotions, and I found myself disposed to abandon my mission as one of an almost hopeless character, owing to the passions and ignorance of the people. There were four murders within the first six weeks after my arrival in the immediate neighbourhood of Ralahine, all marked by features of great barbarity, and all had relation to ejectments from farms, exclusion from occupancy, or in competition for land, which was, in fact, a battle for life and the means to live.

CHAPTER IX.

ORGANISATION OF THE NEW SYSTEM.

The character of the people on the estate and in the neighbourhood of Ralahine may, in some measure, be estimated by the conditions existing in 1831, at the time of my arrival. It was not easy to convince them of the advantages of united homes and combined social arrangements. The peasantry, although living in extreme wretchedness, from their irregular employment and small earnings, were strongly attached to their old customs and isolated, miserable cabins, with their apparent freedom. But their poverty and necessitous circumstances still rendered them the slaves of ignorance, vicious habits, and inveterate prejudices.

The advantages of dwelling in comfortable, well-built houses were pointed out—the convenience and economy to be obtained from cooking and other domestic arrangements were explained—the advantages to be derived from regular employment all the year round, under their own direction and control, were stated, and that at the end of their first year's crop the net profits, if any, after paying rent and interest, would be divided among all. The loss of time, labour, and money involved in the isolated dwelling, where one person is occupied in cooking and other misdirected labour was not appreciated.

It was intended to abolish the custom of individuals owning and rearing animals. It is well known that the pig in Ireland often pays the rent of the poor cottier, and, as a consequence, has, in some districts, an amount of freedom by no means favourable to sanitary conditions necessary to health. It was intended to abolish this private property in animals as far as was deemed practicable and desirable. Owing to the prejudices of the people, it was necessary to adopt arrangements so as to train them to the system, while their individual independence was left as free as possible, except in one or two matters deemed essential to the future harmony and prosperity of the community.

At the time when the proprietor had prepared a draft of the agreement, as to the rent of the estate and the capital invested, and I had drawn out the regulations for the government of the society, the peasantry were in the highest degree of excitement. They were alike insensible to their political and religious leaders, and indifferent to the terrors of the law, acted in defiance of increased military force, and continued to perpetrate the most savage acts of vengeance and barbarous deeds of violence. At this stage it was determined to urge forward the completion of the requisite premises, houses, dining-room, lecture-hall, store-rooms and dormitories. The old baronial castle was also made available for temporary occupancy. The people who had been employed on the estate were called together on the 7th of November, and the objects of the association explained, as stated in the preamble, to be the following :—

Objects of the Association.

1. The acquisition of a common capital.
2. The mutual assurance of its members against the evils of poverty, sickness, infirmity and old age.
3. The attainment of a greater share of the comforts of life than the working classes now possess.
4. The mental and moral improvement of its adult members, and
5. The education of their children.

THE ADVANTAGES WHICH THE PROPRIETOR ANTICIPATED WERE :—

1. To obtain a higher rent for his land.
2. Better interest for his capital.
3. To secure the punctual payment of these.
4. Security for the advances made upon the labour of the people.
5. The safety of the stock, machinery, and capital entrusted to their management.
6. To effect these objects in accordance with the laws, and, at the same time, improve the condition of every member of the association.

It was soon evident that there was an absence of agreement as to the advantages offered by the " new system." I was told that some of the men were under the impression that the plan would be broken up in a very short space of time. Owing to the state of excitement in the district, we were, from motives of personal safety, compelled to begin with those employed on the estate. Had we had full power to freely select the members, some of those first admitted would have been rejected. Where the first members are intelligent and agreed as to their principles and aims, few regulations will be required. But with a population such as I found in County Clare, it was necessary to make regulations in accordance with the objects in view and the conditions of the people.

The spirit of opposition, which had been partly smothered by the decisive measures adopted to organise the associattion, now manifested itself, and it became a struggle between those who were favourable for the experiment to proceed and those opposed to this section, which should be in the ascendant in influence and numbers. To restrain the open manifestation of party feeling, I suggested to Mr. Vandeleur that the whole of the adult members should be subject to the ordeal of the ballot as a silent, critical, estimate of each before he submitted the rules and the agreement for adoption and signature. Before carrying out this trying ordeal, the proprietor made the following address to the people :—

My Friends,—Before I give the rules for adoption I find it necessary to have each member submitted to the ordeal of the ballot, because I have reason to suppose that some persons of one class have an unkind feeling towards others of another class and are also opposed to the system. I could now, before the rules are adopted, or the agreement signed, turn out any or every person that I supposed was not cordially inclined to co-operate for the benefit of each and all. So that now the only terms on which I will allow the rules to be passed will be that each person on the list, according as they are alphabetically arranged, shall be balloted for, and if any person should unfortunately happen to be rejected by a majority of the persons voting, I must, however disagreeable to my feelings, dispense with the services of that individual ; and I can not any longer suffer that person to continue in the establishment. I am aware that some strangers are here, but Mr. Craig expresses a strong desire to be elected or rejected by ballot, as it would be more agreeable to him to be admitted by the general body than that you should have it to say that I forced him on you. Then, when the members are elected, we will adopt the rules and sign the agreement ; and I trust that those who now oppose the society will find it their interest to carry the rules into effect.

Mr. Vandeleur thought that at that time the ordeal of the ballot might prove somewhat hazardous to myself.

The ballot was very satisfactory in awakening self-examination; none were rejected, while the mutual criticisms had the desired effect on the members. The result was seen in the confidence given to those who were disposed to give the system a trial.

The members admitted were:—

Adult single men	21	
Married men	7	—28
Single women	5	
Married women	7	—12
Total		40
Orphans under seventeen years of age:—		
Boys	4	
Girls	3	— 7
Infants under nine years		5*
Total		52

* A Professor of one of the colleges at Cambridge tells me he a had, few days ago, a conversation with a prominent member of Mr. Gladstone's Cabinet in reference to the history of Ralahine, who expressed his surprise at the small number of children in comparison with the number of members. It was said the proportion of children was much greater in Ireland. The observation is a just one, and our explanation is simple. The paucity of infants was not the result of any economic design. We had seven married couples. One couple was advanced in life, and had left their children self-independent outside the society. The herdsman and his wife (the dairy-woman) were somewhat exceptional, as they never had any children. Widow Murphy brought six, three of whom were adults and three youths. There were five infants to five married couples. That children were not deemed an impediment was indicated in the fact that we accepted three orphan boys, under twelve years of age, besides the widow's children. At a later period the members elected a labourer, with his wife and their four young children, because the parents were known to be hard and industrious workers, and the children would more than cover the cost of their maintenance before they were full members at seventeen years of age. If the society began with few infants, Irishwomen are marvellously prolific. Owing, probably, to healthy surroundings, and a vegetable diet, they had very easy confinements, and were

When the members had been elected the rules were submitted for their adoption or alteration, with the exception of those relating to spirits and smoking, gaming, and expelling, which it was deemed necessary to make permanent laws for the benefit of the members. The regulations were found to be adapted to the objects in view, and exercised a healthful influence in promoting harmony and good feeling in the community. The admission of new members by ballot, after a week's trial, worked satisfactorily. No new members could be admitted without a written consent from the president. The candidates made their application through the secretary, which gave me the opportunity of aiding in the selection of applicants likely to meet the requirements of our position. As already stated, none sought admission for the first few weeks, but in about two months we got into very satisfactory working order. At first, the members could scarcely realise the full advantages of their independent position.

In the law No. 10 it was expressly stated, and marked with italics, that no individual was to act as steward. The members when employed as labourers had been subjected to a great deal of harsh and despotic treatment. They had been ordered to their work without any regard for their feelings. Great mistakes were often made in sending too many to one part of the farm and too

---

soon out and at work again. We were in a fair way to correct the figures in future, and the probability seemed to be that at some not distant day we should have to swarm off to a new hive or enlarge the old one. It will be seen by the regulations that I had anticipated some difficulty under this head, and a rule was made to bar marriages, which was proved to be effective in two instances. Two members had to leave the society, because the other members, by ballot, refused to admit the parties selected, who, in both instances, were servants in the hall of the landlord and president.

few to another; and much time, passionate language, and bad feelings were manifested in the arrangement of details. All the affairs of the society were now under the control of

THE COMMITTEE.

The appointments of the members to work were made by the committee, which met every evening to arrange the business for the following day. Every member had a number, as well as every horse and implement, and the labour directions were written on slates, ruled and numbered, to correspond with the members' numbers, and at the close of the committee's business the slates were hung in conspicuous places on the wall in the public dining-room, each individual examined the labour directions, to ascertain the kind of labour he was to perform the following day; and those who had witnessed the disorder, the waste of time and labour, the bad feelings, the violent language, personal insults, and vulgar oaths used in the appointment of the men to work before the society began, would have been astonished at the change which was accomplished by this arrangement. In the morning all went quietly to the business appointed for them, without any commands or complaint. As all violent passion is a waste of life force, the peace and quietude of the feelings and emotions were results of the highest satisfaction and a recompense for all my labour and anxieties.

Should the weather or any other unforeseen circumstance require some individual alterations during the day, the members of the committee acted as sub-committees, made appointments to suit the emergency, and thereby prevented a deal of time and labour from being wasted. If the general committee at any time made any injudicious appointment, or overlooked any branch which it was desirable to have commenced or completed, every adult member was at liberty to make suggestions in a book provided for that purpose, and which always lay open in the committee-room.

It was my duty as secretary to record the suggestions, and from the position I held among them, to see the instructions of the committee carried into effect; and I always endeavoured to convey my remarks and suggestions in a spirit of kindness and conciliation, knowing, however erroneously individuals might act, a spirit of sympathy and charity would ultimately, in connection with proper arrangements, effect better and more satisfactory results than the iron rule of despotism to which they had been so long accustomed; they had, indeed, like many others, suffered so much annoyance from individual caprice that the very name of steward was odious to them. On one occasion, I was requested to act in the capacity of steward; but I declined, for I was deeply interested in the problem as to whether the capital, farm, stock, &c., would be safely managed by ploughmen and labourers. I had provided a "suggestion book," on the plan I adopted in Manchester, in which any member could make any recommendation he deemed advisable. The "suggestion book" was read aloud previous to the appointments being made by the committee for the following day. The plan was of great practical utility, both in the regulation of the work and improvement of the members in the development of capacity and character. The contents of the "suggestion book," and the answers of the committee if "approved," "postponed," or otherwise, were also read aloud by me, and subjected to criticism, at the weekly meetings of the members, who found that a reasonable attention had been paid to their suggestions. A healthy public opinion was formed and brought to bear on the progress of the society. Sometimes very judicious suggestions would be made by men who, all their lives previously, had been treated as utterly unworthy of a moment's consideration. Formerly they were despised by the steward, and treated with indifference, and therefore withheld their suggestions and advice. They received their orders and performed as little work as they could for the small return they obtained for their labour. In the new arrangements, every member felt he had an interest in preserving the property and increasing the produce. The difference in the effect on the manners and deportment of the men was very remarkable. Under a despotic domineering task-master they often appeared sullen, depressed, and dissatisfied. When appointed to their labour by the men they had elected, they were free, cheerful, and contented. The change was from that of slavery to that of freedom. Nothing is more painful to a high moral and generous mind than that of being forced to do the work of a severe unfeeling foreman or employer. Competition is the fruitful source of misery; it corrodes and cankers the mind, and engenders the worst passions of human nature, as egotism and selfishness often govern those who are placed in the position of task-masters. Want and starvation compel men to submit to overbearing treatment, which in the "new system" could not possibly exist; perfect equality prevailed.

In some accounts which have been published of our proceedings at Ralahine, I am represented as acting as steward. This is a misapprehension of my position and motives of action. My aim was to make every member a steward, and, as stated, a law was adopted to prevent the possibility of any single member acting with this despotic power over others.

The great change which had been effected in the conduct of the labourers was soon observed by the people in the neighbourhood. One of the tenant farmers near Ralahine stated that we "had made so great a change in the Terry Alts (discontented labouring peasants) we had amongst us, that he would be glad if we could take a lot more of them."

CHAPTER X.

CONSTITUTION AND LAWS.

The organisation of the society having been determined, it became necessary to consider the constitution and laws by which the members were to be governed, and the following rules having been drawn up by me, and approved by the proprietor, they were submitted to the members and signed. No alterations were found necessary during the experiment. Owing to the disturbed condition of the neighbourhood, I deemed it advisable that the president should retain the veto on the new candidates for admission.

He did not wish to have the responsibility of rejecting an applicant, but the plan was found to exercise a salutary influence, while the prohibition of spirits was made absolute and unalterable by the members.

LAWS OF THE RALAHINE AGRICULTURAL AND MANUFACTURING CO-OPERATIVE SOCIETY.

Preamble.

Objects.

The objects of this association are :—

I. The acquisition of a common capital. II. The mutual assurance of its members against the evils of poverty, sickness, infirmity, and old age. III. The attainment of a greater share of the comforts of life than the working class now possess. IV. The mental and moral improvement of its adult members. V. The education of their children.

Association and Agreement.

1. For the attainment of the foregoing objects, the persons who have signed these rules agree to associate together, and to rent the lands, buildings, manufactories, machinery, &c., of Ralahine, from Mr. John Scott Vandeleur, according to agreement; * and they each of them, jointly and severally, bind themselves to obey the following rules, and to use every means in their power to cause them to be observed:—

Property, when purchased, to be held in Common.

2. That all the stock, implements of husbandry, and other property belong to and are the property of Mr. Vandeleur until the society accumulates sufficient to pay for them; they then become the joint property of the society.

Owner's power of Dismissal.

3. That Mr. Vandeleur have power, during the first twelve months after the society is formed, to direct that any member misbehaving himself or herself be discharged.

Withdrawals.

4. That any member wishing to withdraw from the society have full liberty to do so, by giving a week's notice thereof to the committee.

New Members.

5. If it be found that there is not a sufficient number of persons in the society to carry on the different branches of agricultural and manufacturing industry in a proper manner, one member shall be at liberty to propose, and another member to second, the nomination of a new member; that new member having been approved by Mr. Vandeleur also, may then come upon trial for one week, during which time he or she shall receive diet and lodging only; at the expiration of the week,

---

* This agreement will be found in the next chapter.

he or she shall be balloted for, the majority of members to decide whether such person will be a fit member.

President's Veto.

6. That Mr. Vandeleur be president of the society and of the committee; the committee to nominate a substitute during that gentleman's unavoidable absence.

Appointment of Officers.

7. That Mr. Vandeleur choose the secretary, treasurer, and storekeeper, and that the two former always sit on committee by virtue of their office.*

Visitors.

8. Strangers wishing to visit or inspect any of the departments of this society must request permission to do so from the president or secretary, who shall appoint a member to accompany the party through the establishment.

PRODUCTION.

Each for all.

9. We engage that whatever talents we may individually possess, whether mental or muscular, agricultural, manufacturing, or scientific, shall be directed to the benefit of all, as well by their immediate exercise in all necessary occupations as by communicating our knowledge to each other, and particularly to the young.

No Steward.

10. That, as far as can be reduced to practice, each individual shall assist in agricultural operations, particularly in harvest, it being fully understood that no individual is to act as steward, but all are to work.

---

* This was a salutary precaution on the part of the proprietor, seeing that the whole property entrusted to the association was under their care and in their keeping. A portion of their salaries was defrayed by the society, and the remainder by Mr. Vandeleur, thus making them the servants of both parties; and, as members of the society, they were entitled to share in its profits.

Youths to learn Trades and Agriculture.

11. That all the youths, male or female, do engage to learn some useful trade, together with agriculture and gardening, between the ages of nine and seventeen years.

Arrangement of Work.

12. That the committee meet every evening, to arrange the business for the following day.

Hours of Labour.

13. That the hours of labour be from six in the morning till six in the evening, in summer, and from daybreak till dusk in winter, with the intermission of one hour for dinner.

Wages, and where to be Expended.

14. That each agricultural labouring man shall receive eightpence, and every woman fivepence per day for their labour,‡ which it is expected will be laid out at the store in provisions, or any other article the society may produce or keep there; any other articles may be purchased elsewhere.

Particular kind of labour not to be Compulsory.

15. That no member be expected to perform any service or work but such as is agreeable to his or her feelings, or they are able to perform; but if any member thinks that any other member is not usefully employing his or her time, it is his or her duty to report it to the committee, whose duty it will be to bring that member's conduct before a general meeting, who shall have power, if necessary to expel that useless member.

Distribution and Domestic Economy.

Services to be performed by Youths.

16. That all the services usually performed by servants be performed by the youth of both sexes under the age of seventeen years, either by rotation, or choice.

---

† These were the ordinary wages of the country. The secretary, storekeeper, smiths, joiners, and a few others received something more; the excess being borne by the proprietor.

Children's maintenance borne by the Society.

17. That the expenses of the children's food, clothing, washing, lodging, and education be paid out of the common funds of the society, from the time they are weaned till they arrive at the age of seventeen, when they shall be eligible to become members.

Where Charge to be Made.

18. That a charge be made for the food and clothing, &c., of those children trained by their parents, and residing in their dwelling-houses.*

Public Rooms.

19. That no charge be made for fuel used in the public rooms.

Fuel.

20. That each person occupying a house, or cooking and comsuming their victuals therein, must pay for the fuel used.†

Cooking.

21. That it shall be a special object for the sub-committee of domestic economy, or the superintendent of that department, to ascertain, and put in practice, the best and most economical methods of preparing and cooking the food.

Laundry Work.

22. That all the washing be done together in the public wash-house; the expenses of soap, labour, fuel, &c., to be equally borne by all the adult members.

Sick Fund.

23. That each member pay the sum of one halfpenny out of every shilling received as wages, to form a fund

* These two laws, 17 and 18, were framed for the purpose of securing a superior training of the children by persons specially appointed; and they were effectual.

† Again the laws 19 and 20 were for the encouragement of taking meals at a *table-d'hote*, where better meals were provided at a saving of expenditure; and the charge for turf to those occupying cottages was to stimulate proper economy in its use.

to be placed in the hands of the committee, who shall pay the wages out of this fund of any member who may fall sick or meet with an accident.‡

Wilful Damage.

24. Any damage done by a member to the stock, implements, houses, or any other property belonging to the society, to be made good and stopped out of the wages of the individual, unless the damage be satisfactorily accounted for to the committee.

EDUCATION AND FORMATION OF CHARACTER, &c.

Young Orphans Adopted.

25. We guarantee each other, that the young children of any person dying, whilst a member of this society, shall be equally protected, educated, and cherished with the children of the living members, and entitled, when they arrive at the age of seventeen, to all the privileges of members.

Freedom of Religion.

26. That each individual shall enjoy perfect liberty of conscience and freedom in the expression of opinions, and in religious worship.

Charity of Conduct.

27. That we each observe the utmost kindness, forbearance, and charity for all who may differ from us in opinions.

No Nicknames.

28. That we particularly observe never to call any member by any other than the name he or she may be entered in the books of the society.

No gaming.

29. That no gaming of any kind be practised by any member of the society.

---

‡ There were no sick during the continuance of the society and this fund was discontinued.

All Animals to belong to the Society.

30. That no individual shall keep a four-footed beast, or poultry of any kind.

No Intoxicating Drinks or Tobacco.

31. That no spirituous liquors of any kind, tobacco, or snuff be kept in the store, or on the premises, and any member permitting any of these articles to be brought in, or knowing them to be used and not giving information thereof to the committee, shall be treated as for a breach of Rule 36.

Arbitration.

32. That if any of us should unfortunately have a dispute with any other person, we agree to abide by the decision of a majority of the members, or any persons to whom the matter in question may be by them referred.

Marriage to a Member.

33. That any person wishing to marry another, do sign a declaration to that effect one week previous to the marriage taking place, and that immediate preparations be made for the erection, or fitting-up, of a suitable dwelling-house for their reception.

Marriage to Non Member.

34. That any person wishing to marry another person, not a member, shall sign a declaration according to the last rule; the person not a member shall then be balloted for, and if rejected, both must leave the society.

Travelling and Reception of Freinds.

35. That each member have opportunities and facilities for communications with their relations and friends, and for travelling, and that such time be allowed for these purposes as shall be compatible with the interests of the society and the pleasure of the individual; suitable accommodations shall also be made for the visiting friends of the committee.

Dismissal.

36. That if the conduct of any member be found

injurious to the well-being of the society, the committee shall explain to him or her in what respect his or her conduct has been injurious; and if the said member shall continue to transgress the rules, such member shall be brought before a general meeting, called for the purpose, and if the complaint be substantiates, three-fourths of the members present shall have power to expel, by ballot, such refractory member.

Government, &c.

Ballot for Committee.

37. The society to be governed and its business transacted by a committee of nine members, to be chosen half-yearly, by ballot, by all the adult male and female members; the ballot lists to contain at least four of the last committee.

Departmental Occupation.

38. That the business of the society be divided into the following departments:—1. Agriculture and gardening. 2. Manufactures and trades. 3. Commerical transactions. 4. Domestic economy. 5. Education.

Sub Committees of Superintendence.

39. That for the superintendence of these several departments, the committee shall appoint sub-committees from their own number, or from the members of the society, who shall give reports to the general committee when required.

Minute Book.

40. The committee to meet every evening, and their transactions to be regularly entered into a minute book, the recapitulation of which is to be given at the society's general meeting by the secretary.

Weekly General Meeting.

41. That there be a general weekly meeting of the society; that the treasurer's accounts be audited by the committee and read over to the society; that the "Suggestion Book" be also read at this meeting.

Account Books.

42. The books and accounts of the society to be open for the inspection of all the members.

Half yearly Meetings.

43. A half-yearly general meeting shall be held on the first of May and the first of November, for the election of officers (as by Rule 37), and for the transaction of such other business as may come before the meetings.

Alteration of Rules.

44. That any of the foregoing rules and regulations —with the exception of those relating to the agreement with Mr. Vandeleur—may be rescinded, altered, or additions made thereto, with the consent of three-fourths of the members in general meeting assembled for this purpose.

RULES OF THE INFANT SCHOOL

Parents' Duties.

1. Each child to be brought by its mother to the school at six o'clock in the morning, thoroughly cleaned, combed, and washed.

Hours of Attendance.

2. To remain at school till six o'clock in the evening, during which time the children are to be under the entire care of the mistress, under the superintendence of Mr. Vandeleur, or such other person as he may appoint in his absence.

Discipline.

3. The parents are not to interfere with the discipline of the school, or to take any of the children away, on any account, during those hours.

Constant Superintendence.

4. These rules apply to Sundays as well as other days.

These were the rules at the commencement of the school, but arrangements were afterwards made for the children (not infants) to sleep in the dormitory attached to the school, under the care of the mistress, thereby saving mothers the time and attention required in washing and dressing their children (not always done in the best manner) before going to their labour. It also obviated the necessity of the children rising so early, and prevented all that bustle, irritation of feeling, scolding, and often worse, on the part of the mothers—who had two conflicting duties to perform—to the serious detriment of character of mother and child.

In the playground attached to the school a series of simple instruments were arranged for the amusement and exercise of the children without taxing their strength too much, or exposing them to danger or accidents. In unfavourable weather the exercises could to a great extent be performed under cover.

CHAPTER XI.

AGREEMENT BETWEEN LANDLORD AND TENANTS.

The constitution and laws of the society having been voted, we next proceeded to the adoption of the following agreement, as between landlord and tenants :—

Memorandum of an agreement entered into, this 10th day of November, 1831, between John Scott Vandeleur, of the first part, and John Hastings, John Hogan, and Edward Thomas Craig, in trust for the use of the "Ralahine Agricultural and Manufacturing Co-operative Association," of the second part.

The said John Scott Vandeleur lets the lands of Ralahine now in his possession (excepting the paddock, and the wood and Muckinagh, but including the bogs of Ralahine and Dereen), to the said persons, for the use of said society for twelve months, from 1st November instant, on condition that the rules adopted and passed this day are adhered to by the society, and that they deliver the hereinafter-mentioned articles and produce to the said John S. Vandeleur, as soon as possible after they are demanded from the society by him, to be delivered either at Ralahine, Bunratty, Clare, or Limerick, as he may require, free of expense :—320 barrels of wheat ; 240 ditto of barley ; 50 ditto of oats ; 10 cwt. of butter ; 30 cwt. of pork ; 70 cwt. of beef. The society to supply him with hay at 30s. per ton, and straw in return for the manure, and dung for the garden at 1s. per load ; or, in case Mr. Vandeleur wishes, he may take labour notes in lieu of the above articles, at the rates above mentioned, which nltes are to be received in the store in payment for articles, at the same rate as they are sold to any of the members.

The land to be cultivated on the most improved system, as arranged by the committee. The buildings to be kept in the best order and repair, and all the roads, fences, and houses that the committee may order, are to be made in the best and most permanent manner. No tree to be cut down without leave of the said J. S. Vandeleur and for each tree that he permits to be cut down, two are to be planted in some place appointed by him. Neither back nor front lawn to be cultivated or pastured, without leave of said J. S. Vandeleur.

If, on the 1st of November, 1832, on making up the year's accounts, it appears that the produce above-mentioned has been delivered by the society to the said John S. Vandeleur; that all the wages have been paid to the members; that the society owe no money to any person whatever; that all these agreements have been performed; that the same quantity of farm produce is on hand then that there is now; that the same quantity of farm work is done then that there is done now; that the number, and quality, and value of the stock and implements, as per inventory, has not decreased—then it will be competent for the committee, at a general meeting of the society, to order the wages of men receiving only eightpence to be raised to tenpence and the wages of the women to be raised to sixpence per day for the ensuing year; then if any balance or profit appears to have been made by the exertions of the society, any member wishing may depart from the society, and may take his share of such profit; but no member remaining in the society shall have any power to draw his proportion of the profit.

All the rest to Accumulate.

All the profits to accumulate until the value of the stock and implements, as per inventory, is paid off, when the interest on the same will cease to be paid to the said J. S. Vandeleur, and the society will be at liberty, at a general meeting, to decide what is to be done with the accumulated profits.

In case of deterioration, the farm to revert to landlord.

But if, on the 1st of November, 1832, the society does not appear to have prospered; that they have not delivered the amount of produce agreed upon herein to the said J. S. Vandeleur; that they have not paid the wages of the members; that they have not the same quantity of produce on hand and work done as is done now; that the amount and value of stock of all kinds on hand does not amount to the same value as the amount now delivered, as per inventory—then the whole concern, crops, implements, stock (live and dead), returns into the hands of the said John Scott Vandeleur, who may then make such a selection of the members only as he may think proper, and as he thinks are willing, and understand how to carry on the "new system;" or to act with his property in any manner he may think proper.

Unused Water Power.

It is also agreed upon that if the water-power in Dereen is not converted to some use before twelve months, that it ceases to be included in the property handed over to the society.

Certain Laws of the Society Unalterable.

It is also agreed upon between the said parties that the society have no power to change or alter rules—No. 31, relative to spirits and smoking; No. 29, to gaming; No. 36, to expelling.

JOHN SCOTT VANDELEUR.
JOHN HASTINGS.
JOHN HOGAN.
EDWARD THOMAS CRAIG.

Signed by the above parties this 10th day of November, 1831

Although not specified in the agreement, the society was allowed to build additional dwellings as the members increased, on the condition that for every tree growing on the estate cut down for building purposes two young trees should be planted. There was abundance of limestone for building purposes, and plenty of turf for converting it into lime for mortar. The proprietor defrayed the expense of glazing and slating, the members performing all other requisite work in construction. A large lake supplied abundance of excellent water for domestic use, and a fall of about 30-horse power for driving machinery for manufacturing and other purposes. The estate was in every respect admirably adapted for the objects in view.

CHAPTER XII.

NO NEW APPLICATIONS FOR ADMISSION.

Among the members first admitted at Ralahine there were only eighteen efficient labourers. Additional aid was therefore highly requisite, but during the first few weeks no applicants offered themselves as members of the "new system," as they termed the association. At first the prejudice against the society was very strong, as it was held to be a sort of poor "House of Industry." Labourers who were only employed regularly about three months in the year preferred their accustomed poverty to regular employment, plenty of food and clothing, hours of recreation and enjoyment, free from a tyrannical task-master or steward, with an equal share in the net profits, if any, at the end of the year, to the conditions which their prejudices fancied the system would throw over them. One of the families on the estate admitted at first, consisted of a widow and six children. The old woman was haggard and care-worn, and three of her children were under age, one decrepit, and herself only able to look after the poultry. Yet she literally wept on leaving her wretched cabin, which soon tumbled into ruins.

It will be seen that the mode of paying the rent differed from the old and accustomed methods. The prices ruling the Limerick markets in 1830–31 were taken as standard prices during the existence of the association, for the six articles in which rent was paid, and it was felt to be just, and gave satisfaction to both landlord and tenants. If the produce of the farm had increased, or say doubled, temporarily by the effect of an exceptional season, or permanently by improvements on the part of the members, the society would have appropriated the difference. In the case of permanent improvements, the landlord would, of course, have been benefited by an increased value of the property. In either case the increase would have arisen from causes beyond the control and quite independently of the landlord. It would have arisen either from increased industry, care, and skill, improved modes of tillage, increase of the acreage under cultivation, or from an unusually favourable action of nature's laws, affecting a greater absorption of the various elements of the earth and atmosphere, which go to form plant life. The proprietor would not have supplied this extra industry, these improved methods, or extra forces in nature's laws. To give the increase or double quantity to him would have been unjust. On the other hand, suppose the market price had fallen one-half under the usual relations between landlord and tenant, the community must have given double quantities, or their equivalents, to make up the £900 for rent and interest. This we conceived would have been inequitable, since the prices of farm produce are often ruled by parties and influences over which the society could exercise no control. Prices are often governed by gambling speculations among men whose interest it is to cause fluctuations in market values. Prices are also often seriously affected by the state of the currency under false monetary laws, and the purchase, sale, and price of bullion in the money market. The interest at the bank at one time will be double that at another period. The interest of £100 will be equal to £200 by the action of the bullionists, thereby enriching the capitalist at the expense of the producers. But these are conditions quite beyond the control of the society, and independent of the labour, skill, and prudence of the members. To make rental of land vary with the fluctuations of the money market, under an unscientific system of exchange, gives undue and very unfair advantages to capital, and keeps the producing classes in the condition they are in at present, the mere slaves of toil ; that condition being aggravated by the injustice arising out of a false system of exchanges, making the drones richer by the abstraction of greater profits and the producers poorer.

Under the arrangement made with the proprietor, the society had the full benefit of the skill, industry, and enterprise of its members, and had the advantage of good seasons ; while, on the other hand, the landlord reaped the advantage of any advance in market prices, which in course of time would result from increased demand. Had the society neglected the proper cultivation of the land, it would have risked having no surplus to divide among the members, and have been liable to risk the loss of its occupancy.

It may be urged, as it often was, that the plan is not consistent with the recognised rules of political economy. This science is progressive, and when justice and happiness are included in its aims, it will embrace many social questions now ignored. What the relations of the landlord, the tenant farmer, and the labourer are at present all the world knows. What the labourer might become in peace, prosperity, and happiness Ralahine has demonstrated under the most adverse circumstances.

Although agriculture has made vast strides, and machinery has given immense facilities for the production of manufactures, the position of the labourer both in England and Ireland is worse physically than it was four or five hundred years ago. When the labourer received a penny a day, he could buy a sheep with a week's wages. It would have taken a month to purchase one in England, and a much longer time in Ireland, when we began operations. In the middle of last century wages, after many fluctuations, settled down to about a bushel and a half of wheat. In the middle of the present century a farm labourer could purchase for his eight shillings, received for a week's work, only one bushel of wheat. In the fourteenth century a week's wages would buy four bushels of wheat.

The arrangements with the landlord at Ralahine were of such a nature as to preclude all anxiety on the part of members as to prices. The great object was to crop as large a portion of the land as possible, so that the community might reap the benefit of any excess over and above the fixed quantity of produce agreed to be delivered as and for rent and interest of capital invested and advanced by the proprietor. The arrangement was considered so fair, and the quantities were so regularly delivered to his agents in Limerick, or the other nearest market, and the cash was paid to his credit so promptly when required, that no cause of dispute ever arose. One or two of the members thought the bargain somewhat too favourable for the landlord; but to these even, the constant employment and the many social and domestic comforts enjoyed by association in the community out-weighed all objections on that ground. Had the rent been £100 less it would have made a vast improvement in the condition of the members.

The landlord admitted that the rent was too high. It was very evident, also, that if the land had been the property of the members they would soon have become very prospersous and wealthy.

CHAPTER XIII.

COMFORT ATTRACTS ATTENTION.

Our proceedings, however, at Ralahine soon began to attract attention, owing to the comfort, freedom, and industry of the members. Some who were incredulous made pilgrimages to see the real state of the people, and to enable them to form an opinion by the light of their own eyes and observations. One large farmer found it difficult to understand how a large estate of 618 acres, with 268 under tillage, could be carried on except under the despotic, driving force of a steward or manager; and, in order to satisfy his curiosity, came to Ralahine, and happened to find one of the members at work, and alone, under the following circumstances:—

The water-course which supplied the power for the thrashing mill, as it left the lake on the estate, passed under the old mail road between Limerick and Ennis; and near the tunnel the masonry had given way and obstructed the flow of the stream. The visitor was surprised to find one of the members standing up to his middle in the water, repairing the wall, and entered into conversation with him to the following effect:—

VISITOR—Are you working by yourself?

MEMBER—Yes, sir.

V.—Where is your steward?

M.—We have no steward.

V.—Who sent you, then, to this kind of work?

M.—The committee.

V.—What committee? Who are the committee?

M.—Some of the members, sir.

V.—What members do you mean?

M.—The members of the new system—the ploughmen and labourers.

The gentleman subsequently expressed his astonishment at finding this solitary workman so industrious and executing the work so well, under the novel circumstances.

It doubtless seemed marvellous that the men who a few months previously were known to have been in a wild state of agrarian turbulence and insurrection, and engaged in, or conniving at, outrages of the most fiendish character and of the blackest hue, were now peaceable, industrious, and contented; and engaged in the severest kind of labour without any steward or manager to direct or coerce them, and cheerfully performing their duties as directed by a committee of their own selection.

Some of the objectors, at a distance, deemed the system delusive, because no steward was appointed, and especially so as the members were mostly Terry Alts, ignorant and unthrifty labourers, who, if they did not quarrel among themselves, would blunder from want of knowledge and experience, or probably, on finding themselves their own masters, would be tempted to idleness or over-indulgence, and so make a fool's paradise, and eat themselves out of house and home; others objected to the system because it was not in accordance with the established rules of political economy and the relation of landlord, farmer, and labourer, and the three profits or rents.

One of the causes which operated in producing the change may be understood from the reply of one of the members to another visitor, Mr. John Finch, a merchant of Liverpool, a shrewd, practical man of business, who remained three days with us, examining all the details, which, with the rules, and agreement, with which I supplied him, he published in his report in fourteen letters in a Liverpool journal, and had intended to publish a second series had he lived a short time longer. He made both a private and public appeal to me to make known the results as he saw them. His warm and enthusiastic admiration of our proceedings was the result of a close and searching investigation of the arrangements. He had travelled repeatedly as a merchant through Great Britain and Ireland, mixing intimately with the people, knowing them well, their virtues and their weakness, so that his testimony may be considered an impartial and valuable one. Mr. Finch, among many other incidents, relates that: "A sensible agricultural labourer with whom I conversed when at Ralahine, in contrasting their present with their former condition, under a steward, said to me, 'We formerly had no interest, either in doing a great deal of work, doing it well, or in suggesting improvements, as all the advantage and all the praise were given to a tyrannical task-master, for his attention and watchfulness. We were looked upon as merely machines, and his business was to keep us in motion; for this reason it took the time of three or four of us to watch him, and, when he was fairly out of sight, you may depend we did not hurt ourselves by too much labour; but now that our interest and our duty are made to be the same we have no need of a steward at all.'"

He thus declares his opinion of our proceedings in one of his published statements: "To me it seems impossible to devise arrangements more easy, practical, and economical than those adopted at Ralahine. How paltry, mean, and despicable are all our poor laws, charity institutions, and even our national colleges when compared with these simple, rational, and natural proceedings."

Mr. Finch gave the following testimony in his evidence before a committee of the House of Commons, appointed in 1834, to inquire into the causes, effects, and best means of preventing drunkenness. In reply to a question as to the amusements then prevalent among the people, he said: "I saw an agricultural institution (Ralahine) in Ireland, last year, in the County Clare, where they had dances two evenings in the week generally, but always one. I was present at one of these dances, and the ploughboys and labourers, and their wives and daughters, had a musician, and spent a very merry evening, without a single drop of intoxicating drink of any kind. All the arrangements and laws of this institution are so excellent—they point out so clearly the certain means of removing immediately and for ever ignorance, mendicity, pauperism, drunkenness, and crime that exist in both countries, without any extra outlay of capital, or interference with existing institutions, either in Church or State, that I am determined to devote a considerable portion of my time to the promulgation of them, and should be most happy to have an opportunity of stating them before Parliament, or before a select committee of either House. They are most important to landholders."

Among the visitors who took a deep interest in our proceedings was the brother of Archbishop Trench, who wrote a very fair and discriminating account of his visit, and sent the paper for our criticism and approval.

Not wishing to attract attention during the first year's operations, he was requested not to publish it.

The change effected by the operations of the society was so great and manifest that its influence extended to the neighbourhood, and seemed to have a magical effect. The comparative independence of the members became a subject of honest pride and boast of the humblest among them, and a subject of astonishment and wonder, not only in the neighbourhood and the county, but in the whole of that part of Ireland. The "new system" was the subject of conversation at their places of assembly far and wide. The peasantry began to hope and indulge in the expectation that other landlords would adopt similar arrangements on their estates. After the society was in complete and satisfactory working condition agrarian outrages began to cease, and while murders were increasing in Queen's County and other districts they were unknown in County Clare during the existence of the society; and none occurred on the Ralahine property for more than thirty years afterwards. At the present moment County Clare is a proclaimed district under the Coercion Act of 1881.

The influence of the society was manifested in various ways. The Government had done nothing but apply the terrors of the law and make military displays of force. The landlords had issued a circular asking "what was the best thing to be done?" The priests formed a committee with the hope of subduing agitation and repressing illegal movements, but abandoned the plan and dissolved the organisation. The happy conditions at Ralahine now attracted all eyes and awakened hopes in the hearts of the people.

The Secretary for Ireland, the Hon. E. G. Stanley,

the late Earl of Derby, in a speech in the House of Commons in favour of a Coercion Bill for other counties, pointed to the County of Clare, and attributed the then peaceful condition of the people to the "Peace Preservation Act." This was a great mistake, and an utter perversion of historic truth. The landlords had done nothing, the clergy were powerless, while the police and the military had been daringly defied. The only change effected in the condition of the peasantry was what was occurring at Ralahine. A few simple arrangements which we had introduced had made a complete change in the character and ideas of the people. The men were no longer slaves to their passions; no longer subject to the harrassing control of a man imposed on them. Sons of the slaves of toil, and ever bearing their chains, betwixt cringing servility and revolt. they were now free men, subject to no laws but those which, having contributed to make, they willingly accepted; working with their equals, owing obedience only to the committee whom a majority of the members had temporarily invested with authority, the development of character was very striking in the altered conditions. Men who had hitherto been sullen, moody, and discontented, were free, frank, and communicative.

Our progress and happy proceedings became widely known, and for a time a source of trouble and inconvenience. There were no poor-laws at that date, and the paupers were legion. When a labourer died, his wife and family were thrown upon charity, and with a wallet and a tin can went on tramp, and many of these, as well as professed mendicants, visited the members, and implored charity. "For the love of the holy Mother of God," they craved "a bit or a sup," milk or potatoes.

It was, however, soon found that the members had nothing they could give away as private property, except labour notes, which were useless to any but the members.

The fame of our proceedings had been published in a short notice in the *Times* and other newspapers, and a well-educated youth, named Joseph Cox, employed in a solicitor's office, in London, resolved to visit the society, and in a spirit of enthusiasm walked to Liverpool, crossed the sea to Dublin, and walked thence to Ralahine, and became so pleased with our proceedings that he made an earnest appeal to the president to be admitted as an associate, and with what result will be seen in the sequel. It was a mistaken kindness to place a slim, well educated youth, who had never handled any instrument heavier than a lawyer's pen, in competition with stalwart Irish labourers whose whole life had been one of toil and industrial effort with the *Slahn*, or spade. Experienced working agricultural labourers only, and under intelligent guidance, should be the pioneers in co-operative farming.

CHAPTER XIV.

A MARKET ON THE SPOT.

In estimating the condition of the members at Ralahine, it was very evident to all visitors that there was a vast improvement in the relations of the people with each other, and with their obligations as industrial producers. There was great freedom and independence, for there was no despotic steward to domineer and browbeat them. No man was "master"; none were treated as servants, or slaves, to carry out the behests of an imperious task-master. All were on an equality with relation to the committee, whom they had themselves appointed or elected for the purpose of directing the practical working details. Each and all were mutual helpers for the benefit of all.

But although there was this inestimable privilege of each being master of himself, as sovereign of his own individualism, it is necessary to observe that for the mutual advantage of each and all, order, system, and arrangement were necessary for the success and economy of the practical working operations of the community. To give to the members their due influence, plans were adopted calculated for enabling all the members to exercise their legitimate influence over the operations of the community, and the arrangements of the committee, who met every evening to appoint the members, when necessary, to their required duties on the following day. This had a very marked educational influence in arresting attention, prompting thought, awakening reflection, and in developing character.

Each member had a number engraved, as previously stated, on a ruled slate, opposite which the special work to which he was appointed was recorded. The book slate was on view in the public hall, so that all could learn, without any harsh or austere command, the special instructions of the committee, who appointed themselves on the same plan. If a member decided on being absent, which was rarely the case, the fact was indicated by omitting the registry of work on the

DAILY LABOUR SHEET.

The appointments of the committee were copied on a large sheet, ruled so as to show the work done during the week, and to indicate whether executed on the "Farm," for the "Family," or for "Improvements." The Daily Labour Sheet was exhibited during the following week on the wall of the lecture hall, so that every member could see how far the committee were judicious in their regulations. As the plan of the Daily Labour Sheet was useful, it may be suggestive to those who desire to adopt the preparatory arrangements necessary to secure order, practical prudence, and security in the incipient stages of a new community. The following gives an illustration of a portion of the sheet, with an abstract of the expenditure during the same period :—

Daily Labour Sheet, from Monday the 9th to Saturday the 14th day of January, 1832.

No. of Members.	Name.	Monday, 9th. Farm	Family	Improvements	Weather dry. Employment.	Tuesday, 10th. Farm	Family	Improvements	cold and wet. Employment.	Days at Rate paid each	Farm	Family	Improvements	Cost weekly per Farm.	Family.	Improvements.	Total.
										d.				£ s. d.	£ s. d.	£ s. d.	£ s. d.
1	T. Carmody	1	..	..	Ploughing ..	1	..	..	Carting manure ..	10	6	..	..	0 5 0	....	....	0 5 0
2	J. Davine ..	1	..	..	Steaming potatoes	1	..	..	Steaming potatoes	8	6	..	..	0 4 0	....	....	0 4 0
3	Anne Davine	..	1	..	Dormitories ..	..	1	..	Dormitories ..	5	..	6	..	....	0 2 6	....	0 2 6
4	T. Euright	..	..	1	Cottage wall ..	..	..	1	Cottage wall ..	8	4	..	2	0 2 8	....	0 1 4	0 4 0
5	C. E. ..	..	1	..	Washing	..	1	..	Washing	5	..	6	..	....	0 2 6	....	0 2 6
6	P. F. ..	1	..	..	Herding	1	..	..	Herding	8	6	..	..	0 4 0	....	....	0 4 0
7	W. F. ..	1	..	..	Carting turnips ..	1	..	..	Trenching wheat ..	8	6	..	..	0 4 0	....	....	0 4 0
8	G. M. ..	1	..	..	Trenching ..	1	..	..	Trenching ..	8	6	..	..	0 4 0	....	....	0 4 0
*	* * * *																
50	T. W. ..	1	..	..	Trenching :	1	..	..	Trenching ..	8	6	..	..	0 4 4	....	....	0 4 0
									Farm					7 4 2	0 15 8	2 4 4	10 4 2
									Family					0 15 8			
									Improvements ..					2 4 4			
														10 4 2			

ABSTRACT OF LABOUR SHEET.

For Week ending Saturday, January 14th, 1832.

FARM EXPENDITURE.

	£	s.	d.	£	s.	d.
Carting out and mixing compost manure	1	1	6			
Ploughing in Granahan and Calf-field	0	13	4			
Washing and steaming potatoes ...	0	6	0			
Thrashing and preparing wheat for market	0	18	0			
Conveying wheat to Limerick ...	0	10	8			
Pulling and drawing in turnips ...	0	8	0			
Trenching in wheat	0	9	0			
Attending and foddering cattle ...	0	16	0			
Carpenter's labour on farm ...	0	7	4			
Smith's ditto	0	4	8			
Herding stock	0	5	0			
Dairy	0	5	0			
Manufacturing wool into frieze ...	0	3	2			
Poultry	0	2	6			
Sundries to farm	0	2	0			
Attending and feeding pigs ...	0	4	0			
Superintendence, education, and accounts	0	8	0			
				7	4	2

FAMILY EXPENDITURE.

	£	s.	d.	£	s.	d.
Attending dining-rooms	0	2	6			
Steaming potatoes and other vegetables for family	0	2	0			
Attendance on dormitories	0	1	8			
Sewing and preparing beds ...	0	0	10			
Washing clothes	0	5	0			
Infant schoolmistress	0	2	6			
Sundries to family	0	1	2			
				0	15	8

IMPROVEMENTS EXPENDITURE.

Carpenter's labour	0	10	8			
Smith's ditto	0	6	0			
Attending slater	0	2	0			
Storekeepers	0	16	0			
Clerk of accounts	0	6	8			
				2	4	4
Total cost of labour for fifty members for one week				£10	4	2

E. T. CRAIG, *Secretary*.

WEEKLY EXPENDITURE OF MEMBERS.

By reference to the ledger accounts for this perod, it appears that the consumption of produce, &c., during the week ending January 14, for the entire community, of fifty adults and seventeen children, was as follows :—

	£	s.	d
Milk, 202 quarts	0	16	10
Potatoes, Vegetabes, and Fruit	2	0	6
Butter, 13¾ lbs.	0	9	2
Mutton, 9½ lbs.	0	3	2
Eggs, 32	0	0	8
Fuel for married members	0	0	7½
Lodgings for single, and rent of houses for married members	0	10	0
	4	0	11½
Balance left with members for clothes, savings, &c.	6	3	2½
	£10	4	2

The members received these balances in labour notes. The payments for the work done amounted to £10 4s. 2d.; but owing to the great economy and superiority of the associative arrangements over the isolated, individual, and wasteful competitive system of society, the community had a market on the spot for a considerable portion of their own produce, and thus returned nearly one-half of their drafts on account of labour, which were on the same scale as wages in the neighbourhood. The balance in the hands of the members, to be saved or used at their discretion, was an important consideration when they had still the prospect of an increment in the profits arising after paying the landlord the amount of produce as rent and interest for the use of his capital. The net profits, after rent and interest were paid, were to be divided among the members in addition to their wages. Our plan was exceedingly simple, safe, and practical, and, when contrasted with the conditions which prevail, affords very great advantages in the economy of time, labour, and money over the present wasteful methods of distribution, improvidence, competition, and social antagonisms. The vast waste of time and capital in the carriage, for mere purposes of distribution of articles of daily consumption in contrast with the potential social arrangements of enlarged homes in the centres of fertile lands for the production of agricultural produce, is a great loss of labour and a sad waste of human life. There is, for instance, wasteful improvidence in half-a-dozen milk carts and the same number of bakers, greengrocers, and coal dealers rushing to and fro in the same street, where one of each kind would meet all the requirements of the people. Here are twenty-four horses and men employed to do what four of the former and the same number of the latter could supply. Thus twenty men and twenty horses more than necessary have to be maintained at the expense of the consumers in a very limited district. The folly and absurdity of this waste would be laughable if it were not so serious; the result of ignorance, and want of organisation and method. The present modes of distribution are at variance with the simplest rules of economy, as well as the cause of a vast amount of misapplied toil, waste of capital, social wrong, tyranny, and injustice.

At Ralahine the dairy was not a hundred yards from the dining-room and cottages. As the result of a market on the spot, new milk from the cow was sold at 1d. per quart, and potatoes at 2d. per stone.

Mr. John Finch, in detailing the results of his three days' visit to Ralahine, gives the following

COMPARISON OF RALAHINE LABOURERS WITH OTHERS EARNING THE SAME NOMINAL WAGE.

"England's curse, and Ireland's bane, intoxicating drinks, formed no part of the diet at Ralahine; it consisted principally of an unlimited supply of new milk, and a great abundance and great variety of vegetable food of their own growing, of the best quality, and in great perfection, and no man or woman in England or Ireland, that I have seen, lived cheaper, laboured harder, or looked better than they did. And let those who find fault with the milk and vegetable diet of Ralahine reflect that 'no human being has any natural right to require another to do that for him or her, that he or she would refuse to do for that individual: *and, therefore, in a rational state of society, all those who will eat beef, mutton, veal, and pork, must in turn kill the animals for themselves.*'

"The following is an account of the provisions delivered to the society, consisting of eighty-one persons, men, women, and children, for their consumption, in the week ending 6th October, 1833, being more than an average of their weekly expenditure :—

		£	s.	d.
446 quarts of new milk at 1d.	...	1	17	2
Potatoes and other vegetables	...	2	13	6
Butter, 12s. 1d. ; pork, 19s. 7½d.	...	1	11	8½
Cottage rent, 4s. 3d. ; turf for fuel, 9d.		0	5	0
Food and lodging for 81 persons	...	£6	7	4½

Averaging less than 1s. 7d. per week each.

"Let the condition of the inmates of Ralahine now be compared, as it ought to be, with the condition of labourers whose earnings are the same—the hand-loom cotton-weavers earning in 1833, by hard labour, from 2s. to 5s. per week; the Staffordshire nailers earning from 4s. to 10s. ; the agricultural labourers in the south and east, with their 6s. to 9s. per week ; the labourer in the north and east of Ireland badly employed, and working for from 8d. to 1s. per day. But more particularly compare their condition with the wretched peasantry in the south of Ireland (where this experiment was made), without any employment at all for months in the winter, whose average earnings throughout the year will not amount to 6d. per day, and this for the support, not of the individual only, but of a large family—his wife and children having no other employment than attending to a little plot of potato ground, and begging—their only food a scanty supply of potatoes, and those frequently of bad quality—the whole family covered with filth and clothed with rags—all herding and sleeping in the same apartment (upon a bit of dirty straw) with pigs, ducks, poultry, dogs, goats, cows, and asses, on an earth floor, covered with mire and manure of every description—

the apartment a miserable mud cabin, roofed with thatch or sods, often without window or chimney, and nine times out of ten with a dung heap before the door —no bed, no furniture, scarcely any utensil save a cast-iron pot, in which the potatoes are boiled for the family dinner, and in which afterwards the surplus and refuse are served up to the grunters. Let them compare the condition of these unfortunate individuals with that of members of this society, and remember that Ralahine was only the first step from such a state as this; they will then be able or form a just estimate of its value."

WEEKLY MEETINGS OF MEMBERS.

At the weekly meetings of the community the minutes of the committee and their votes in the suggestion book were discussed, and appointments as recorded on the labour sheet were criticised.

It was objected that too much labour was sometimes directed to improvements instead of to the farm. Once or twice the conduct of members became subject to discussion. The meetings and friendly criticisms had an educational influence over the whole of the members. Mr. Finch records the fact that what he calls a trial happened while he was on his visit of inspection. He states the affair thus:—

"No litigation either before a lawyer or magistrate was allowed; all disputes were settled by arbitration among themselves. A passionate man called another a bad name, which led to a quarrel, a trial for which happened while I was there. This gave me an opportunity of witnessing a meeting of the Ralahine parliament. Twenty-nine adult males and seventeen females were assembled on this occasion, and the case was entered into, discussed with all due seriousness, and ended in a caution to the offenders against a repetition of such conduct."

The weekly meetings had the happy result of interesting the whole of the members in the proceedings of the committee and in the success of the new system. The views of the committee recorded and read from the "Suggestion Book" were discussed, and the practical value of certain methods of dealing with the land was also canvassed, and formed an excellent basis for the educational training of the people, and the formation of correct notions and higher phases of character than is possible with the usual isolated methods of dealing with the labourer and the artisan. The apathy and indifference of the agricultural labour under the present narrow, selfish, and unfair methods of absorbing all the profits instead of sharing them with the producer, is the natural result of the system which excludes him from a share of his own earnings. To tell the half-fed and ill-clad labourer that the interest of the workman is the same as that of the master is to offer an insult to his understanding and to common sense.

CHAPTER XV.

THE IRISH LABOURER AND THE IRISH LAND BILL.

The difficulties in the way of organising the association of the labourers were great and peculiar, probably greater than ever existed in any other experiment of the kind. And when the obstacles arising from the ignorance and the prejudices of the first members were overcome, no fresh applications, as before stated, were made during the first three months by persons outside for admission as members. The number of labourers was too few for the extent of land under tillage,and more were required. The poverty, destitution, and wretchedness of the people around the estate were very great and depressing, as it was impossible to relieve them.

In visiting some of the miserable cabins at a little distance from Ralahine, I found the occupants in a most deplorable condition. In one mud cabin, on the hillside, I found a poor widow with four young children, without any means of subsistence, except what they could obtain by begging, on tramp, through the district. The only furniture consisted of a bed of straw, spread over a wattled foundation on the floor of clay, with the stump of a tree for a seat, and an iron pot for cooking. The poor woman was accustomed to smoking a dirty clay pipe to deaden her sensibilities, and allay the pangs of hunger, thereby soothing her cares at the cost of a halfpenny a day. It was justly remarked by Mr. Villiers Stuart, M.P., that "No one who has not travelled in Ireland could understand the condition of the farm labourers' dwellings, with their floors of mud, rotten thatch, and the one chamber, which did duty for drawing-room, parlour, stable, and pigsty, with the sole accommodation of one bed, in which people of all ages and both sexes had to lie."

A return in the House of Lords has recently been issued at the instance of one of the peers,* showing the total area of grass lands, including meadows and pastures, and the total area of tillage lands including all kinds of crops and fallows, in Ireland for the years 1855, 1860, 1865, 1870, 1875, and 1880. In 1855 there were 10,872,968 statute acres under grass, and 4,436,035 of tillage lands; in 1860 the numbers were—grass, 11,078,152, tillage 4,411,454; in 1865, grass, 11,501,905, tillage 3,998,968; in 1870, grass 11,740,706, tillage 3,885,945; in 1875, grass 12,354,005, tillage 3,399,001; in 1880, grass 12,168,933, tillage 3,186,665. This return clearly shows that while the area under grass has increased more than a million and a quarter of acres, the amount of tillage —the great source of labour, employment, and increase of cattle food and farm produce—has diminshed in an equal proportion.

The increase of grass implies a reduction of food supplies and less demand for labour. A reduction in tillage betrays a lack of confidence, an absence of capital, a diminished labour market, with an increase of destitution, poverty, and agrarian crime. An increase in small cotter holdings will not increase the area of permanent profitable employment. It is the absence of employment which forces the Irish cotter to visit England during harvest time; and when agriculture is in a prosperous condition

* Lord Cloncurry.

each labourer will return with £8, £10, or £12 saved from his earnings. This competition with the English peasantry is an evil to both countries, and demonstrates the truth of the statement that it is regular employment which is required, but which small holdings cannot supply. This lack of regular employment has long been the source of Ireland's difficulty and discontent. A report of a select committee of the House of Commons, on "The Employment of the Poor in Ireland," states that "Those districts in the south and west presented the remarkable example of possessing a surplus of food whilst the inhabitants were suffering from actual want. The calamity may, therefore, be said to have proceeded less from the want of food itself than from the adequate means of purchasing, or, in other words, from the want of employment. An intelligent agriculturist, who visited Ireland, alleges that a large portion of the peasantry live in a state of misery of which he could have formed no conception, not imagining that any human beings could exist in such wretchedness. Their cabins scarcely contain an article that can be called furniture; in some families there are no such things as bed clothes, the peasants strewed some fern, and a quantity of straw over it, upon which they slept in their working clothes; yet, whenever they had a meal of potatoes, they were cheerful. The greater part, he understood, drink nothing but water. This statement appears confirmed by the testimony of many of the witnesses examined by your committee, who agree not only in this melancholy description of the condition of a considerable portion of the Irish peasanty, but agree also in attributing it to the total want of employment in which they are left. In some parts of the country, one-half of the entire population is stated to be without employment; in others, the proportion is said to be still greater; and all the witnesses examined agree in attributing, to a considerable degree, the turbulent spirit of the peasantry and their excesses to this cause." As it was then, so it continues. Want, poverty, and turbulence are the accompaniment and twin brothers of want of employment. Ralahine points to the remedy.

I have seen the habitations of the peasantry of France, Belgium, Holland, Germany, and Switzerland, but in no country have I seen such wretchedness and misery as I saw in the cabins of the peasantry in the south of Ireland.

The social evils of Ireland arise out of combined causes, such as the uncertainty of enjoying the fruits of labour from rack-rented holdings; the poverty and ignorance of the peasantry in relation to the science of agriculture; the absence of capital, and the isolation of individualism in farming operations. To advocate small holdings and peasant proprietors in Ireland is a great mistake in political and social economy. It is a retrograde movement, as unwise as the advocacy of abandoning the facilities of railway rapid progress for a return to the single packhorse of the olden time, or the disuse of the forces of nature, and the substitution for the spinning-jenny of a thousand strings of Penelope's distaff and single thread.

Although there were no applications from labourers to become members of our community at the commencement, we had as I have already mentioned a very great number of visits from tramps and miserable mendicants. It was evident that some of the married women in the society, in accordance with their accustomed habits had supplied the applicants with vegetables and milk obtained from the common dining-room. The result was what might have been expected. The place was the centre of attraction to a class of persons who, however deserving they might be of charitable assistance, it was not desirable to encourage at Ralahine. The disposition to indiscriminate charity was soon checked, while it served to illustrate the practical wisdom of the plan adopted of paying wages in "labour notes."

In organising the society it was in contemplation to place all the members on an equality as to the means of enjoyment, and the landlord to give whatever extra payments were required by the skilled labourer and artisan. But as the peasantry were ignorant of the principles of our social union, and could not clearly comprehend the advantages to be derived without having immediate control over their income, it was decided to give the usual nominal wages, and so regulate prices and domestic arrangements that the members would become prepared for a higher social condition in the course of time.

The money value of the Irish labourer's work varied according to the price of his provisions. His wages were at that time uniform at 8d. or 10d. per day. The price of potatoes, however, varied from 2d. to 8d. per stone. But wages were not advanced because provisions were more costly. The political economist may urge that the law of supply and demand must be taken into consideration. This is true in the present irrational relations between labour, land, and capital. The object should be to obtain a rule of justice, if we seek the law of righteousness. This can only be fully realised in that equality arising out of a community of property, where the labour of one member is valued at the same rate as that of another member, and labour is exchanged for labour. It was not possible to attain to this condition of equality at Ralahine, but we made such arrangements as would impart a feeling of security, fairness, and justice to all. The prices of provisions were fixed and uniform. A labourer was charged 1s. a week for as many vegetables and as much fruit as he chose to consume; milk was 1d. per quart; beef and mutton, 4d. and pork 2½d. per pound. The married members occupying separate dwellings were charged 6d. per week for rent, and 2d. for fuel.

CHAPTER XVI.

LABOUR NOTES.

The notes were printed on stiff cardboard, about the size of a gentleman's ordinary address card, being two-and-a-half inches long by one-and-a-half wide. The unit note represented the value of one day's labour, thus :—

8 PENCE
FOR ONE DAY'S LABOUR.

There were other notes representing a half, a quarter, an eighth, and a sixteenth of a day's labour respectively, or the equivalents for fourpence, twopence, one penny and a halfpenny. The usual wages of a labourer in the district at that time was the foundation of the rates affixed as value. Above the unit notes were others, representing one shilling, or one day and a-half's labour, printed in red ink for distinction, and another equal to three day's labour, or two shillings. These notes were received as currency at the store for articles produced on the farm, for materials for clothing, fuel, rent, &c. The labour was recorded daily on a "labour sheet," which was exposed to view during the following week. The members could work or not at their own discretion. If no work, no record, and therefore no pay. Practically the arrangement was of great use. There were no idlers, and as the currency was only available within the community, the members gradually found themselves in possession of ample funds after their current expenses were discharged.

At the first issue of the labour notes there was some objection to them on the ground that "they were not money." But when it was found that on application to me they could be exchanged for hard coin when necessary, or for food, clothing, &c., at their own store, the objection died away, and ultimately the notes were preferred to cash payments.

The adoption of a distinctive medium of exchange of this kind had many advantages. It made the working operations independent of the outside ordinary currency. It gave the members full control over their own resources, and prompted them to habits of prudence and economy. The labour notes were not current outside of the community. All the articles provided in the store were genuine. There could not be any motive for adulteration, false weight, or measure. It was a rule that neither tobacco nor spirits were to be sold in the store.

The plan of currency is admirably adapted for associations in isolated positions. Had the experiment at Ralahine proceeded it would have been found expedient to adopt a plan of labour notes less capable of imitation, with sufficient security for the issue, and for cancelling them when withdrawn. As it was, the amounts were correct, and the balance had to be met by means which I had to provide out of my private resources. But this is a contingency that should never be allowed possible, as it opens the way to irreparable wrong and injustice to the labourer. The total amount required for circulation was £50, which should have been secured on the land, or by securities in the hands of trustees.

The advantages of the labour notes were soon evident in the savings of the members. They had no anxiety as to employment, wages, or the price of provisions. Each could partake of as much vegetable food as he or she could desire. The expenses of the children from infancy, for food or education, were provided for out of the common fund. Peasants, who, at the commencement, although young, appeared aged, careworn, and half-famished, soon exhibited a marked improvement in their appearance, becoming healthy, lithe, cheerful, and active. Some that were in rags or poorly clad had two suits of clothes and money saved in labour notes. All these results were attained while paying nominally the same wages as the farmers outside, owing to the economy of our arrangements, while paying an annual rent of £900 for 618 acres, only 268 of which were under tillage and fairly chargeable as being occupied.

The very marked and visible improvement of the members began to be appreciated by the peasantry. Applications were occasionally made by persons in the neighbourhood to become members of the "new system." Each applicant had to submit to a week's residence on trial, and to the ballot on admission. Great attention was paid to the physical and moral characteristics of the new admissions, and the results were highly satisfactory. The applicants had to apply to me, and having due regard to the future, a selection was made of those whose heritage indicated health, vigour, and powers of endurance, The fine, manly appearance of the members often arrested the attention and commanded the approbation of visitors. It is a remarkable fact that the young men who were known before the society commenced, to have been violent Terry Alts or White Boys, proved to be the most active, steady, and industrious members.

In less than six months after our establishment, the moral aspect of the peasantry in our neighbourhood had become changed. No agrarian acts of violence were anywhere known around Ralahine. The peace of the district was so far secure that in the summer evenings of 1832 I repeatedly found small dancing parties held in lanes, branching off from a main road. As evidence of the great change that had occurred among the peasantry towards myself, occasionally on passing some of the cabins and dancing parties young girls would come towards me, and, with smiling faces, curtsy, as evidence of their respect, and as an invitation to join them in their Irish dance. We had dances twice a week at Ralahine, which I always attended, when possible, as thereby amusement was afforded, without the evil accompaniments attendant on the free use of "potheen." I never observed any evil effects from these social gatherings. Young men who had been hard at work in trenching or in following the plough appeared to derive some enjoyment or benefit from the recreation, which always terminated at ten o'clock, when the gates were closed.

It was intended to raise the taste and extend the sources of enjoyment of the members, but there was a danger in attempting to advance too far or too fast for the mental condition of the people, The confidence of the people in my desire to benefit them was shown in various ways. They would sometimes confide in me so far as to relate some of the devices that had been adopted to frighten certain middlemen, landlords, and stewards to force them to convert grazing land into tillage farms, in the dark and desperate days before the "new system" at Ralahine was organised. I was shown disguises used in the neighbourhood, while some of the stories made the blood run cold, and the heart sick at the relation. I was told how the steward of Ralahine was murdered, and why it was resolved to destroy him, and how the man who shot him escaped and was never brought to justice.

The change which had been effected was very remarkable. Less than twelve months had elapsed since the starving and indignant peasantry were united in making what was called a "hell upon earth," to induce the landlords and their agents to allow them to toil and cultivate the land for their subsistence, and now they were peaceable, industrious, and contented under the "new system" which I had been instrumental in establishing. For many generations there had been much pains and many penalties, with much preaching superadded, to induce the order, prudence, and thrift which now pervailed at Ralahine, and which might be introduced into any district requiring them, to render well-directed labour not only useful to the labourers themselves, but beneficial to the rest of society. The change realised among the people by the new system was such as to astonish all classes. Many who had a long experience in the country, considered the plan of letting the estate, the farm buildings, stock, implements, and giving a participation in the profits to a number of wild, penniless, discontented Irish labourers, some of whom had been concerned in scenes of outrageous violence, and possibly murder itself, as little short of insanity.

The results, however, fully justified the experiment. The landlord was relieved from a vast amount of care and anxiety, and the people were industrious, contented, and happy. Many of the peasantry began to hope that other large landed proprietors would adopt the system. On one occasion, when returning from Limerick, a Catholic priest met me, and insisted on my going to his house, and partaking of refreshment, when he spoke in warm commendation of what I had accomplished, and expressed his high gratification at the change and improvement among the people at Ralahine. A few months previously the priests and the clergy were utterly powerless to quell the riotous proceedings of the peasantry, who were rendered furious by their wretchedness. The transformation of the conditions seemed a mystery to many. Social science was more successful than military force. Socialism could induce poor, ignorant Irish peasants to live in peace and harmony with each other. In their individualism every man was for himself at the expense of every other person. Now, each found his interest promoted by promoting the comfort and happiness of all around him.

Ralahine became as "a city set on a hill," and attracted the attention of men of all classes. Noblemen, landlords, thinkers, and writers visited us to discover how far our system would promise a solution of the difficulties with which Ireland was afflicted. Some objections were urged against the "new system," because it was "not in accordance with the rules of political economy, and the arrangement of landlord, farmer, and labourer, and the three rents." It was because we held political economy imperfect that we had united the farmer and the labourer in the same person, and dispensed with the middleman. Social economy aims to unite capital and labour in the same person. Socialism, where only partially applied, had effected what neither the government, the soldier, the priest, nor the political economist could accomplish; and time and experience were alone required to reconcile it with science and biological truths. It is no longer a theory or Utopia, as in the days of Plato and of More, but a real, tangible verity, waiting only for prudent, practical development and extension by those who comprehend its purpose, and can organise the men and materials necessary to success.

CHAPTER XVII.

STIMULANTS, THE CAUSE OF CRIME, EXCLUDED.

The co-operative farm at Ralahine had not been in successful working order for more than a few weeks when it became very manifest that the greatest benefits, arising from our social union, united efforts, enlarged homes, and social surroundings, were derived by the women, youths, and children. The wretched cabins of the peasantry of Ireland contribute largely toward their discomfort, their bad habits, and their misery. It is impossible for children to be trained so as to exercise their better feelings and develop happy dispositions when reared in homes of poverty, filth, wretchedness, ignorance, and often drunkenness, vice, and crime. It is a law of the mental organisation that the faculties are stimulated by their proper objects of excitement; irritation and opposition make combativeness more active, and repetition increases activity. Passion developes passionate natures, until habit establishes impulsiveness in seeking mere sensual gratification. In drawing up the rules for the society it was decreed that no intoxicating liquors or tobacco should be kept in the store or be introduced on the premises by any of the members. This rule was deemed necessary to the peace, harmony, and health of the members, and was never infringed.

Much has been said and written about the difficulties arising from the sudden abandonment of habits to which the people have become accustomed. I have shown in describing the condition of the peasantry, the inveterate custom of tobacco-smoking, and the violence stimulated by indulgence in whisky, which prevailed among the people. The exclusion of stimulants was made to operate at once, and to be permanent. Experience justifies the adoption of this course, as will be seen as we record the facts. No inconvenience arose from the sudden abandonment of the custom of smoking, nor was any attempt made by any of the members to introduce intoxicating stimulants into the community. The advantages to the women by the banishment of whisky from our midst was soon very evident, for we had no drunken quarrels, so frequent among the poor when under the excitement of their poisonous "potheen," and accursed, burning, and inflaming "fire-water."

In making arrangements for united homes it was necessary to have due regard for the health and character of the children of the community. Heritage is a law almost ignored by the prevailing habits of competitive society. Idiocy, imbecility, and intemperance have a closer relationship than many are aware; while nicotine is a poison which destroys nervous sensibility, and lays the foundation of weakness and special diseases in the sight and digestive organs. The tobacco plant is a weed doubly injurious to the land and to the consumer, while the distillation of alcohol from grain is an improvident waste of capital and labour. A drunkard is the most helpless of animals, and a drunken woman the most pitiable creature in existence. They become burdens upon others, a curse to those they ought to cherish, and a tax on the labour of society. In 1833 the quantity of malt used in Ireland for making beer, whisky, &c., was

1,970,000 bushels. From this year there was a constant increase till 1836, when it amounted to 2,511,000 bushels. As the increased intemperance of the poorer classes kept pace with the increase of population and diminished prosperity, it may be inferred that greater poverty, by increasing wretchedness and despair, is one great cause of increased intemperance; while greater poverty is in turn the natural and inevitable result of increased intemperance. It is a fact worthy of consideration that during the period of temperance in subsequent years there was a great diminution of crime. Murder, shooting, stabbing, assault with intent to murder, solicitation to murder, conspiracy to murder, manslaughter—when all put together—were committed 898 times in the year 1839, 503 times in 1840, and 502 times in 1841. In the two latter years the duty on spirits had decreased about one-third, showing an intimate relation between crime and alcoholic-poison drinking; while its diminution corresponded with less crime, less taxation, and greater prudence and prosperity.

I had seen sufficient of the terrible evils of intemperance to insist on the utter exclusion of whisky from our store and dwellings. By the banishment of the cause of quarrels we had peace in the homes, at the weekly dances, and other festivals. In the establishment of the Infant School and playground the mothers were relieved of a constant source of care and anxiety during the day, whilst they had the satisfaction of knowing that their children were, under the kindly superintendence of a competent teacher, acquiring healthy habits, good dispositions, pleasing manners, and a large amount of really useful knowledge and experience of natural objects around them.

The domestic arrangements were also made so as to relieve the women from everything deserving the name of drudgery in household work, and thus helping to make the hours and days pass pleasantly along. Every married couple had a separate dwelling. The youths had their dormitories and sitting rooms. Only those infants that were not weaned were to sleep at home, except otherwise by arrangement.

Owing to the great extension of fever and cholera around Ralahine in the epidemic of 1832, we devoted especial attention to all the known causes likely to develop susceptible conditions in the members, by the removal and deposit in the earth of all garbage. Another means was the erection of an improved ash-closet. As the cleansing of a closet used for personal convenience was a disagreeable office, we devised a plan which raised the closet and seat some eight or nine feet above the foundation; and by placing a flag at a sharp angle, the deposits could fall into a box on wheels, and the soil be covered with fine earth or ashes, so as to prevent the escape of deleterious sewage gases, which are the chief cause of poisonous atmospheres, and the direct origin of typhoid and other fevers analogous to smallpox and cholera.

As evidence of the physical advantages arising from our arrangement to secure the health and comfort of the people, it was a remarkable fact that we had not a single case of sickness or death during the existence of the society, although during the prevalence of the epidemic of 1832, the people were carried off in scores by fever and cholera in the neighbourhood and around us. Limerick seemed like a city deserted from the ravages of the plague and the footfalls of death. Men were afraid to hold intercourse with their kindred lest they should carry away from them the seeds of disease and death. The perfect health and freedom from disease of our members were the results of combined causes–sanitary precautions, nourishing diet, cleanliness, and the happy cheerfulness and contentment arising out of our social surroundings, dances , and weekly amusements. Famishing, discontented, irritable, and miserable people are short lived, while contentment and cheerfulness are essential elements to health and length of life. Hence the great and imperative necessity that the producing classes should understand the system we established at Ralahine for participation in results, and never rest satisfied till they have organised some means for realising similar, if not superior, social surroundings, and where they can share both in the profits and in the management of the wealth they create by their combined intelligence and industry. The system of participation in the management and sharing in the profits transcends all other methods in promoting the social advancement of the producer.

CHAPTER XIX.

LAND UNDER TILLAGE.

The land of the Ralahine estate was of varied quality. Some portions were of stiff clay, requiring a good supply of manure to develop their latent fertility. A considerable portion rested on limestone rocks, which in some places, cropped out on the surface, and rendered the land stony. There was little sand or gravel. A large portion of the pasture land and plantations was capable of being converted into tillage. The bog lands were only fit for fuel, in consequence of the defective methods adopted in cutting the peat for fuel in separate portions without any system of drainage, so that large sections were made bog holes, or pits of water, and the remainder useless for cropping.

The land under tillage was mostly of good quality, and had been cultivated by the plough. The following table shows the extent of the crops cultivated during the second year :—

	Irish.				English.		
	A.	R.	P.		A.	R.	P.
Wheat	40	1	9	...	65	2	0
Barley	14	3	2	...	24	1	0
Oats	20	1	35	...	32	3	8
Potatoes	24	0	0	...	38	3	20
Turnips	19	0	0	...	30	3	4
Mangel-wurzel	4	0	0	...	6	1	36
Vetches	9	3	0	...	15	3	6
Clover and rye grass ...	18	2	25	...	30	0	24
Fallow	14	2	0	...	23	1	27
Total under cultivation and fallow	165	1	31	...	268	0	5

Pasture and plantations ...	172	1	33	...	280	0	7
Bog	39	3	13	...	63	2	36
Orchard	2	1	29	...	3	3	30
Houses	1	1	34	...	2	1	28
Grand Total	381	2	20	...	618	0	26

The orchard was well stocked with fruit trees and vegetables, and the members of the society, on application, had unrestricted use of the produce, and yet two cart loads of apples and pears rotted during one season for want of consumers. In the future a market might have been found, and the planting of fruit trees in the hedgerows would have become highly productive and profitable.

On the northern boundary of the estate, and partly surrounded by woods and hills, was a large lake, which supplied an inexhaustible source of mechanical power, awaiting future use, but running to waste, except what was applied to drive the thrashing machine of eight horse power, a flax scutching mill, a circular saw, and a lathe.

At a short distance here was also water power, equal to twenty horse power, running to waste, which might have become available for the manufacture of frieze, which would have found a ready market at a remunerative price.

It will be seen by the above table that less than one half of the estate was under tillage, so that the rent of £700 per annum made it over £2 per acre, which was held by some of the members as rather too high, and this was admitted by the landlord. Twenty acres of virgin soil, resting on limestone, which always yields very short herbage, was added to the tillage land, by trenching, by spade cultivation, and quarrying by the crowbar the rocks which came to the surface. The stone was utilised in building new dwellings for members. The land, which, in a state of nature, yields only seven tons of grass to the acre, is capable by judicious labour, and the application of suitable manures, of producing from fifty to sixty tons of forage and root crops. Clay soils, however stubborn, barren, and poor, are capable of being made rich and fertile by cultivation ; first by proper drainage, secondly by working into the ground insoluble gritty materials, such as burnt clay, ashes, sand, or fine gravel, to make it porous, so as to admit the air and the rain bearing ammonia, and, lastly, by the application of manure suited to the crops intended to be cultivated.

The new land, which was scarcely equal to feeding a goat, was trenched eighteen inches deep and well manured with stable dung, and sown with potatoes, followed by wheat. Both of these crops were larger and heavier than similar crops on any other part of the farm.

CHAPTER XX.

THE MYSTERY OF PLANT GROWTH.

In co-operative farming, as in other pursuits, experience and skill are necessary to success. Our labourers, although attentive to the cattle, had not the requisite knowledge how to deal with them, and we should have lost several valuable beasts and milch cows if we had not secured the services of an experienced herdsman, and his wife as dairy-woman. The attainment of experience may be acquired without much risk of loss when dealing with inanimate objects, but when organised beings are concerned it is necessary to have the experience at hand to know how to overcome unforeseen difficulties. While the major part of our members were deficient in special knowledge, it was necessary to rely on those who knew particular branches.

Agriculture, in one sense, is an art dependent on observation and experience; in another aspect it is a science where the application of known laws leads to invariable results. The plan of tillage at Ralahine depended on natural manures and a rotation of crops. But science has led to the application of artificial fertilisers, as well as the use of mechanical agencies in place of hand labour and horse work in tillage. Changes have arisen in agricultural practice, and results realised which were never anticipated fifty years ago; lands under the present improved culture yield returns which a few years since would have been deemed impossible. It has been said, with some truth, that the capability of arable land to grow green crops is almost unlimited. The fact is illustrated in the heavy crops raised from market-garden grounds, and the poor soils of Belgium, where the soil is supplied with ample plant food in complete manures, and gives a constant supply of flourishing vegetation.

This progress of invention in tillage by steam machinery opens up to view grand possibilities in the future, Had we had a steel prong digger at Ralahine it would have cultivated the whole 268 acres in about seven weeks, at only seven acres a day ; whereas we kept four ploughmen, with two horses each, at work all the year round, except a few days at harvest, when occupied in leading home the crops. In addition to the ploughmen we had some ten men occupied in spade cultivation. All these men and horses might have been dispensed with in the tillage work of the farm by using the steam cultivator.

Our members at Ralahine accepted with delight the first reaping machine introduced into Ireland, but the half-starved and unemployed labourers outside viewed its introduction unfavourably. As there was some doubt whether the peasantry in the district might not destroy the machine, and to allay public prejudices against machinery—for looms had been previously destroyed when it was attempted to introduce the manufacture of linen at Ralahine—the landlord wrote the following address, which I signed, as secretary, and distributed in the neighbourhood. Its introduction serves the double purpose of showing the confidence of the proprietor in the co-operative principle, in giving the labourers a participation in the profits, and of recording his testimony to the practical success of the system. The experience of Ralahine in using the reaping machine, while others subsequently used were destroyed by the agricultural labourers, shows that if the industrious classes were allowed a just and fair share in the immediate and direct benefits of machinery, they would cease to oppose its introduction into their respective trades.

TO THE AGRICULTURAL LABOURERS OF COUNTY CLARE

Fellow-labourers,—We, the members of the Ralahine Agricultural Association, address you on behalf of ourseelvs and our co-operative brethren, and recommend you seriously to reflect on the evils which the daily improvements in machinery are likely to inflict upon you, unless you adopt some mode by which machinery shall work *with* you, and *for* you, and not against you. All new inventions, such as this reaping machine, instead of injuring, benefit us; for, instead of bending our backs in reaping and squeezing the corn with our hands the whole day, we make the machine do it for us, whilst we have only the comparatively easy operations of binding and stacking. This machine of ours is one of the first machines ever given to the working classes to lighten their labour, and, at the same time, increase their comforts. It does not benefit any one person among us exclusively, nor throw any individual out of employment. Any kind of machinery used for shortening labour—except used in a co-operative society like ours—must tend to lessen wages, and to deprive working men of employment; and finally, either to starve them, force them into some other employment (and then reduce wages in that also), or compel them to emigrate. *Now, if the working classes would cordially and peacefully unite to adopt our system, no power or party could prevent their success.* The plan will injure no one, but place every individual willing to work, with either head or hands, in employment, and enable each society so to train up and usefully educate its children that they could, with ease and pleasure to themselves, create more wealth than they could consume. We have thus united. We distribute what we require of the produce we create in the best and most economic manner, and govern ourselves to our own satisfaction and comfort. Tell the owners of land that if they wish to use machinery beneficially, they should form you into societies, where it cannot injure you, but where you would have an interest in using and protecting it. And should they be induced to unite with you in these arrangements, so advantageous to all parties, they would soon see a great, wonderful, and rapid improvement in the state of the country; there would be no more *starvation in the midst of abundance*, nor any necessity for industrious workmen to leave their homes, friends, and country, for foreign woods and wilds, whilst their native land remains but partially cultivated.—By order of the Committee,

E. T. CRAIG, Secretary.

Ralahine, 21st August, 1833.

The italics are those of the writer of these very remarkable statements, who, as an Irish landlord, had let his lands, buildings, stock, and machinery to me and two others as trustees, on behalf of a number of poor, penniless Irish peasantry, some of whom were discontented Terry Alts, or concerned in certain agrarian outrages, before my arrival at Ralahine. After two years' experience he acknowledges that the "plan will injure no one, but place every individual in employment," and enable him not only to pay his rent, but "create more wealth than he can consume"; and if the owners of land would adopt "these arrangements, so advantageous for all parties, they would soon see a great, wonderful, and rapid improvement in the state of the country." These statements are the utterances of one who was deeply interested in the rental of his land, the safety of his family, the security of his property, and in the prosperity of all around him and they are not more strange than true.

The rigid exclusion of the labourer from the soil will

have unexpected results when it stands out in bold defiance of the usual practice of mankind. Its effects will be most intensely felt when other countries cease to demand our manufactures. Ireland now blooms as a garden and a grave. There are a few immensely rich, surrounded by large demesnes, parks, and pleasure grounds, rich pasture lands, and well-cultivated fields while the vast majority are in extreme indigience, living in miserable cabins, with small pasture lands, little tillage ground beyond the uncertain potato patch, and where bad seasons entail famine, disease, and premature death. The blessings of inventive skill should be held by no monopolising hand, but spread in broad and expanding streams over the whole surface of society. This might be best accomplished by association. This union of capital and labour might be easily effected by the federation of co-operative societies, who could now ensure success in co-operative farming by means of the steam digger, to till the land for grain for the market, or for fodder and roots for the cattle of a dairy farm and a kitchen garden, the produce of which would find a remunerative market in the flourishing stores in the various districts. The purpose of the machine is to save toil, not to supersede the means of subsistence of the labourer By utilising the digger on their own land, under intelligent direction, and sharing the management and the profits with all concerned in the work,co-operators would be realising the lofty ideal of the founders of the movement, and relieve their members from exhausting toil to secure mere bodily wants, and enable each and all to reach a higher culture and lead a better, a nobler, and a happier life.

CHAPTER XXII.

THE DAIRY DEPARTMENT.

One of the most important, interesting, and profitable departments at Ralahine was the live stock, which was estimated at £1,500, for which the association was charged at the rate of £6 per cent. per annum, as interest for the capital. The following list of cattle, in good condition, comprised the whole of the live stock, as it was a rule that all animals kept should be the common property of the society.

LIVE STOCK.

Cows...	37*
Heifers	44
Bullocks	20
Calves	17
Fattening beasts	16
Bull	1
Cattle Total	135
Pigs	51
Sheep	26
Horses	11
Total animals	223

The stock may be divided into groups, as the thirty milch cows, the young stock, and the fattening beasts. All were stall fed, and each group required special food,

* Of these, thirty were milch cows, and supplied the members with milk, and the dairy with butter, which was chiefly sold to Mr. Vandeleur, or paid as part of the rent.

fodder, and treatment. Two men only were required to feed and attend to the cattle.

The value of milk as an article of diet was very forcibly illustrated by its great sustaining power and flesh-forming properties. Our members lived chiefly on vegetables, consisting mainly of potatoes, which could not have sustained the muscular vigour of the men without the milk, of which each had a pint at every meal, making an average of ten quarts per week, but scarcely any animal food. As has been already stated, the men were remarkable for their healthy vigour; all were lithe, robust, strong, and muscular.

STALL FED CATTLE.

Many large farmers and landlords are at the present moment making an attempt to form an association for the "purpose of showing that dairy farming could be carried on at a profit upon arable as well as on grass farms." This question was decided by our dairy at Ralahine, as the cows and fat beasts were all stall fed by roots and fodder raised on the farm. For this purpose we had devoted thirty acres to turnips, thirty to clover and rye grass, nine to vetches, and six to mangel-wurzel. The young stock were occasionally turned out upon pasture land. It would have required more than three times the number of acres to have sustained the same number of cattle; while the produce of milk would have been considerably less, owing to the waste arising from exertion, cold, and wet. Grazing is a great waste of both labour and capital.

CHAPTER XXIII.

THE AIMS OF THE FOUNDERS OF THE CO-OPERATIVE MOVEMENT.

There were no railways in Ireland at the time our "new system" existed at Ralahine, and as we had agreed to make our annual payments for rent in the following items, less or more of each to make up the full value of the produce, as estimated by the prices in Limerick market at the time we commenced, we had to deliver our fat stock to an agent in Liverpool, The landlord took the risk of the variations in the market prices, and the society the risk of good or bad seasons.

PRODUCE TO BE RAISED FOR RENT.

6,400 stones of wheat, at 1s. 6d. per stone		£480
3,840 stones of barley, at 10d. per stone	...	160
480 stones of oats, at 10d. per stone	...	20
70 cwt. of beef, at 40s. per cwt.	...	140
30 cwt. of pork, at 40s. per cwt.	...	60
10 cwt. of butter, at 80s. per cwt.	...	40
		£900

The cereals could be delivered to agents in Limerick, but the fat stock had to be sent to the Liverpool market. We sent sixteen stall-fed fat beasts to Dublin under the care of two herdsmen. They were driven along the road at a considerable loss of fat. I followed and delivered them to the agent, who paid the money received for them into the bank to the credit of the landlord. This method of paying the rent was a simple one, and gave certainty to the landlord and satisfaction to both parties.

On my return to Ralahine, I found one or two causes of anxiety had arisen from conditions operating outside of the society. Fever and cholera, as I have shown, made it necessary to devote great attention to the sanitary surroundings of the community. Great numbers were dying around us. Fortunately, we had neither sickness nor deaths during the whole of the time of the society's existence, and which I have endeavoured to explain as the result of sanitary precautions.

Although the society had made great progress, there were still one or two features which made it necessary to be cautious in relation to increasing any responsibility as to the comfort and happiness of others. As evidence of our peculiar and transitional condition, I did not feel justified in venturing to take back with me one who subsequently joined the community. The first twelve months were considered experimental, as stated in our agreement. There was not the same prudence on the part of the members. They felt the comfort and security of their position, and while marriage added to their domestic felicity it did not involve the same responsibilities as those which existed outside the society. While the women were employed the wife could save something weekly out of the labour-notes paid as tokens of work done, after discharging the cost of food, fire, and rent. The children when weaned were maintained at the schools, at the expense of the community. Ireland had a population of 8,000,000, and the poverty, wretchedness, and misery which existed had ended in a rebellion which the example and success of our proceedings had arrested and subdued in county Clare and the province of Munster, and encouraged in the Whiteboys and labourers the hope of others adopting the new system.

Viewed from the standpoint of the political economist, our social arrangements in relation to the maintenance of the children was the weakest point in the " new system." In an educational point it was our best feature.

We had, however, made some provision against imprudent marriages with persons outside of the society, in requiring the individual selected by a member to be submitted to the ordeal of the ballot, and if rejected by the members, then both parties were excluded, and the member had to leave the community. This had occurred in the case of a young man, the nephew of the steward who was shot.

During my absence in England three of the members attended at a " wake," where the friends of the young woman who was rejected were also present. Whisky, as usual, did its mischievous work in arousing the passions, which ended in a faction fight, when a stone struck one of the party, causing his death. Our blacksmith, a very active and industrious member, was charged with the act of throwing the stone, convicted, and sentenced to seven years' transportation. His assistant was dismissed from the society, and another member was suspended for being mixed up with the quarrel, and becoming liable to legal proceedings.

In these circumstances we have evidence of the great importance of excluding intoxicating spirits from the store and dwellings within the boundary of the estate. At the commencement of our proceedings we had some difficulty in obtaining members sufficient to cultivate the land. It was now become a matter of life and death to gain admission.

Four of the members married women out of the society who were balloted for and admitted. A young woman who had the care of the younger children married at a later period the chief gardener employed by the proprietor, but he was rejected by the members when submitted to the ballot. These were the only cases that were rejected, but the decisions acted as a salutary restraint on the young against imprudent associations and marriages, and consequently on population. The members became, though unconsciously, instrumental in their own improvement and elevation, as well as in the regulation of their numbers. This is a most important aspect of the population question, especially in the economy of productive labour, and the future political government of Ireland. Political economists with few exceptions, evade the question. They speak of "cheapness," and "competition" as essential to the accumulation of capital. In Ireland they had both competition and cheapness of labour, and yet great wretchedness and poverty. The remedy proposed is emigration. They overlook the fact that this is an evasion of the question, for cheapness, emigration, and population may go hand in hand in other countries and still leave the problem unsettled.

CHAPTER XXIV.

EDUCATION OF ADULT MEMBERS.

The limited education among the adult members at Ralahine has been already indicated, and I found very great difficulties arise in the course of my efforts to supply the deficiency, owing to the fact that the uneducated knew but little of the English language, while the most ignorant were the most superstitious. The largest proportion of words they used were in their native tongue—probably three of Irish to one of English. Owing to the influence and bigotry of religious factions and political prejudices, the education of the Irish peasantry had been disgracefully neglected, and now the English legislature and the peace and order of the country have to pay the penalty of their guilty indifference in the past. They could not give secular knowledge, because the priests would not have it from Protestant religious sources. There was only one Hedge school between Limerick and Ralahine, a distance of fourteen English miles! These schools were of the most wretched kind. One which I visited occupied a mud hovel a short distance from the mail road to Ennis. The cabin had only a very small opening for a window, which afforded no available light for the little ragged urchins assembled under the guidance of a pedagogue almost as ragged as themselves. Some of the younger ones were sitting or lying on the floor; others, a little older, were on benches; while a few tall ones were holding their books to the light from the door. In a corner were several pieces of turf and some potatoes, which each had brought as a school fee for the schoolmaster. Many of the children lived several miles away from this hedge school, and the schoolmaster had to go through Cratloe Wood to his distant home with his heavy and bulky satchel of school fees, paid in kind, as the poverty of the people left them destitute of any other currency.

Although education was lamentably neglected, the people were shrewd and keen observers of political movements. Newspapers were inaccessible, from their high price of sevenpence each, and I never saw one in the hands of the residents of Ralahine.

To educate the young men to a better knowledge of English I adopted the plan of M. Jacotot, by taking an easy chapter of a book, and pointing to the words of the first phrase one by one, requesting the young student to repeat them after me. The next day the first phrase was repeated and two or three words added. This plan was continued for a time, and gave the reader a knowledge of many words in the English language which he could at once pronounce on seeing them in print. The process was a slow one, but effective with one or two, while it was given up by others.

Many visitors who came to examine for themselves the results of our proceedings, which they had either heard described by their friends who had been over the establishment, or had seen some notice in the London *Times* or other newspaper, were often highly gratified with what they saw, and did not withhold their approval, as already stated. One visitor, Lord Wallscourt, was so favourably impressed that he adopted our plans on one hundred acres on his estate in Galway, and with gratifying results.

CHAPTER XXV.

ADMISSION OF NON-AGRICULTURAL MEMBERS.

In framing the rules for the admission of new members to our society, I arranged that each applicant should procure a written application from the president, so that he could exercise his veto both as to the fitness, character, and number of the candidates. Each candidate had to undergo a probation of one week, so that the members might be able to decide as to whether or not the applicant was eligible as an associate, before being subjected to the ordeal of a ballot. The candidates had to apply to me for an introduction to the president, and I had to give advice, which prevented the admission of some Terry Alts, whose antecedents, when I became acquainted with them, showed that organisation and character have a closer relation than is generally admitted.

Experience led me to the conclusion that one week was too brief a time to judge of the moral and mental conditions essential to lasting and satisfactory results. A longer period of novitiate, or trial, is desirable for the sake of the candidates themselves. Two instances illustrate this view, and they are interesting from the fact that they were admitted wholly on the strength of the urgent recommendation of the president, and in opposition to my advice, as neither of them was qualified for the hard and laborious duties which the members had to go through to discharge the heavy rent and interest for the land, stock, and capital advanced.

It is a remarkable fact that these two members, introduced through the kind recommendations of the president, were the only members that retired from the society of their own accord. The fact of their departure was for a time taken up as an argument against our success, until it was made evident that they left of their own free will.

The first instance of a member desiring to leave was that of James M'Guire, who was previously a servant or valet to Mr. Vandeleur. Expressing a wish to become a member, and to assist as a storekeeper, he was elected and appointed to that office, but it was soon evident that his assistance was not required at the store; and as he had not been trained to agricultural labour, he voluntarily proposed to leave and seek a situation suited to his experience and habits.

The other instance of voluntary secession was that of Joseph Cox, to whom reference has been made. His case was interesting from the fact that he was the son of a respectable family in London, had received a good education, and been placed in a lawyer's office. He had, however, been powerfully impressed by the lectures and teachings of the advocates of social progress, in their exposition of the superiority of socialism over the selfishness and evils of competition. The eloquent advocates spoke of conditions that must be the result of long-continued efforts under circumstances favourable to the end in view. He became strongly impressed with the importance of the social movement. A paragraph in the *Times* and other papers recording our progress had arrested his attention, and in a spirit of enthusiasm he resolved, if possible, to join our society. As already stated, he walked from London to Liverpool, and from Dublin to Ralahine. Our society had only been in operation a few weeks on the arrival of the slim youth of about eighteen to visit and examine our proceedings. He was so delighted that he induced Mr. Vandeleur to propose him as a member.

Owing to the conditions under which we had undertaken to rent the estate, stock, and machinery, it became necessary to devote great attention to the crops necessary to meet the demands of the proprietor. These conditions made it imperative that great attention should be devoted to hard and laborious agricultural labour in trenching and spade work. This must be the case in all early efforts in the establishment of co-operative farms. Where there is suitable land, sufficiency of capital under the direction of experienced agriculturalists, the task will be easy and the success certain. But all interested in the concern must be workers and derive direct benefits from the operations. It had been proposed by Mr. Owen that Government should provide the funds necessary for the land, buildings, stock, and machinery. This would involve the absence of freedom and the repression of individual independence. It is far better for social progress that the people economise their resources by securing the profits of distribution and production and unite their savings to build enlarged homes, mansions, and manufactories in healthy surroundings, so as to develop the productiveness of the land as dairy farms and market gardens, thereby finding healthful employment for all who could be trained to the work. The people would then soon find themselves in prosperous conditions, and could govern themselves to their own satisfaction without the restraints of Government tutelage, and the evils arising from political or religious partisanship which would ultimately lead to divisions and separation.

In the circumstances under which we began our

operations it was necessary that every member should be capable of exerting himself in some agricultural labour. The majority of our members were illiterate agricultural labourers, who knew little about the theories and wonders of social science and intellectual progress, but they were capable of long continued labour on the land. All our arrangements were in harmony with the greatest economy, comfort, and social happiness of the members, who soon began to appreciate the benefits of our arrangements.

Although the members felt the advantages arising out of our combined exertions and social arrangements, which seemed to give a new aspect to their daily life, and render existence a real enjoyment, yet they could not appreciate the superior intelligence of the young lawyer; and, as he could neither dig nor plough, they viewed him as a useless member—a drone eating the honey gathered by the workers. This was irksome to Cox. He felt he was unequal to the ignorant but laborious members. He could wield the pen, but could not labour with the spade; he was, therefore, appointed to work which the young boys could execute. True Socialism requires equality of conditions, opportunities, and reciprocity of sentiment. Physically, young Cox became greatly improved. He came among us a slender youth of eighteen years of age, with delicate frame and fingers, and not in vigorous health. A few months' training in the farmyard expanded his chest, gave breadth to his shoulders, and made him strong and robust; but, socially and morally, he felt the lack of sympathy arising from the disparity in intelligence and manners between himself and those around him. The fact that the native Irish language was the principal means of conversational intercourse among the members was another source of dissatisfaction to him. On the other hand, he was equally unsuited to the condition of the people. His superior education raised him beyond the sphere of their usual train of thought; and he soon seemed to lose all hope of finding congeniality among the labourers around him. The enthusiasm with which he had been fired, and which prompted him to leave home in opposition to his friends in London, had been sorely tried by the hard, practical experience of life at Ralahine, and he left after a few months' sojourn among us. His experience shows clearly that enthusiasm without fortitude, faith, and a spirit of endurance, is not sufficient in the early stages of a co-operative farm, and is a suggestive lesson to those who may follow our example. He found himself in the course of a few weeks in an uncongenial atmosphere. The peasants knew nothing about the beauties of literature, the fascinations of the drama, or the mysteries of the quadrille, but they could plough the land, raise the limestone rock, and dance the Irish jig with hilarious enjoyment. Had the young scrivener joined the association at a later period he would have found the society better prepared for his peculiar condition, and his physical weakness and inexperience excused. He gained, however, in form, strength, and practical experience, but his enthusiasm had been rudely shaken; his young dreams were not realised, and yet it was not the fault of the community that he was disappointed. He had been a "dreamer of dreams," in anticipating a condition at Ralahine which it was not possible to realise under the conditions by which we were governed and inflexibly controlled, and yet we had accomplished what people around us considered to be marvellously successful results in improving the people. The stern and arduous work we had to go through was too heavy and too exhaustive for the strength of his will and his powers of resistance.

Persons in the condition of body and mind of Joseph Cox should not seek to join a co-operative farm in its early stage, but wait till the foundations are securely laid, when work, leisure, study, and relaxation will be available and in harmony with the conditions, prosperity, and happiness of all.

The friends of Cox supplied him with means for going to America, and I was glad to hear, subsequently, that he was employed by a respectable merchant at St. John's, at a salary of £100 a year. He had pleasurable recollections of Ralahine, and although he was only with us a few months, a very beneficial improvement had been made in his character in giving him a more practical tone of mind as to the means necessary for advancing the social science of life.

It is only justice to state that he was a striking instance of the advantages resulting from embuing the mind of co-operators with the principles and objects of association for social progress. Those who are thoroughly convinced of the justice of mutually sharing in the products of labour are more earnest, more devoted, and more to be relied on than those who only seek the wages of the mere hireling. It is desirable that the leaders of the co-operative movement should establish distributive stores with reading-rooms, &c., attached, in agricultural villages, so that the members may be systematically educated and embued with the principles involved in mutual co-operation; that men accustomed to agricultural labour and dairy farming may be prepared to devote themselves to the great work of co-operation in connection with the land.

CHAPTER XXVI.

THE INFLUENCE OF CIRCUMSTANCES ON CHARACTER.

The great change that had been effected in the habits and characters of the members was very strikingly shown in the different way in which they regarded the property on the estate under the control of the society. Before the association was formed, the labourers conceived that their own interest was opposed to that of the landlord and would attend to nothing beyond their allotted appointments for the passing moment. If a bullock broke through a fence and trampled down the growing wheat, they would say, "It's no business of ours—let the herdsman see to it." The wheat and other crops were injured and property destroyed because they got neither profit nor thanks for their extra trouble, and sometimes abuse for their interference. They seemed to believe that it was their interest to encourage clandestinely the destruction of property, as it might create a greater demand for their labour. This state of feeling was completely reversed. Not a single article, potato or turf, was wasted, for they discovered that the preservation of the property was really saving their own labour. It is evident that the same motive—self-interest—will produce opposite results when governed by opposite controlling circumstances. What the leaders of humanity have therefore to accomplish is the arrangement of conditions calculated to make self-interest harmonise with the interest, prosperity, and happiness of all. But to be able to accomplish these highly important conditions it is necessary for all to be guided by a knowledge of the laws that govern the feelings, emotions, and mental powers of mankind. A knowledge of technical science alone will fail to realise complete success, although necessary and useful in practical operations. A knowledge of the Science of Society and its educational influence is necessary. Mere physical science may contribute to material progress, while social science is essential to the higher conditions involved in harmonious associations such as were indicated at Ralahine.

Before the existence of our society, as already stated, some of the members were identified with the parties concerned in violent agrarian outrages against obnoxious landlords. The new circumstances in which they were situated gradually acted upon them, and their feelings and conduct became changed. They no longer sought to improve their condition by the destruction of the property of their neighbours. They felt they had a manifest interest in preserving the property which provided them with the means of enjoying constant employment, regular wages, comfortable dwellings, education, and social intercourse, and happiness they had never hoped to enjoy in this world. Even in political matters they became less sympathetic than during the agitation for emancipation and reform. They were more practical. Formerly several attended the great meetings held by O'Connell. Not a single member lost a day by attending at the vast assemblages to hear the "Great Agitator." The habits and hopes of the people seemed to have been quite changed by their improved circumstances, and altered relations with regard to the land and the landlord. They were not like the same men. The most intelligent among them clearly comprehended their position, and the advantages to be secured in sharing in the profits after paying the rent and interest for the capital employed. They saw and felt that their comfort and prosperity were intimately dependent on their success in the tillage of their land. The Clare election gave to O'Connell great influence over the peasantry of the county, and whenever a political meeting was held at Ennis, great numbers were accustomed to attend, and sanguine hopes were awakened among the people that the Repeal of the Union would be secured, to the benefit of Ireland and the Irish. Our members began to see that constant, profitable employment was the true remedy for the prevailing distress, and the new system showed how this could be secured.

THE HUNTERS HELD AT BAY.

The love of excitement by the Irish is well-known, and the hunting season afforded opportunities that were thoroughly enjoyed. It was customary, when the fox took across the lands near where labourers were at work, for all to join in the chase for some distance, and in the excitement would trample down young crops, break down fences, viewing the hunt as capital fun.

A great change was evident when the members of the society began to appreciate their position, as they would not allow anyone to disturb the fences in pursuit of game. The proprietor had a younger brother whose income was too limited to enable him to keep a hunter. He was too proud to seek useful employment, too aristocratic to condescend to work for his maintenance, and he spent much of his time in hunting, with his fowling-piece slung over his shoulder; and on a shaggy pony would sometimes traverse the boundaries of our estate, but the members objected to his riding over the young crops. He was a melancholy illustration of the evils arising from primogeniture, living in a miserable, loose condition of solitary bachelorhood in a small house near the lake.

During the winter of 1832, a hunted fox crossed the mill-course near the rickyard, and took across the orchard and over a seventy-acre field of young wheat, in the best state of tillage. The field of mounted huntsmen were seen galloping towards the farmyard gates, which were solid timber, some eight feet high, fixed on rollers running on rails. By a sudden impulse the members in the farmyard closed the gates, and brought the huntsmen abruptly to bay, and thereby put a stop to the chase for the day. The horsemen were astonished at the presumption of the men of the "new system," who had thus dared to interfere with their sport. The men who composed this wild group, chasing a poor frightened animal, were farmers, tradesmen from the neighbouring towns, clergymen, and "squireens." The latter were the connections or younger sons of respectable or wealthy families, like the brother of the proprietor of Ralahine, having little or no patrimony of their own, but who scorned to demean themselves by any profitable occupation. Many of the huntsmen loudly expressed their disappointment and indignation at the rudeness of the labourers daring to talk to them about "young crops" and "damaged fences!" I had repeatedly heard some of the young members express themselves strongly against the practice of hunting over tillage land. I had given no advice in the matter. Had Mr. Vandeleur been on the spot he would, doubtless, have permitted the horsemen to enjoy their gallop over the wheatfield after the hounds and poor Reynard, who, on that occasion carried his brush home in safety.

HELP FOR THE WIDOW AND THE FATHERLESS.

Although the members had shown their dislike and opposition to the huntsman's pursuit of the fox while destroying the fences and the fruits of their labour, they had much sympathy with the people, and took repeated opportunities of showing it. During the ravages of cholera in 1832, great numbers were laid prostrate and many died. In visiting Limerick the sufferings of the people were painful and depressing—the wailing of women, the consternation and screams of children, and the speechless agony of those afflicted, were sad and sorrowful. The city became almost deserted. Most of the houses in George Street were closed, the owners having fled in terror from the prevailing contagion, while in the poorer districts of the old town the people were also silent and the thoroughfares deserted. The epidemic prevailed also around Ralahine, and many died in the cabins of fever or cholera. As previously stated, we were free from sickness and death. A report reached Ralahine that the crops of a poor widow, who had lost her husband by fever or cholera, would be lost in consequence of the death of him who had sown but was not there to reap, and the absence of means to pay for reapers. On the Sunday following all the young men of the community took their sickles and cheerfully travelled to the desolate home, cut the poor widow's wheat, and harvested her crops free of all cost. This benevolence was shown in other similar cases arising from the harrowing bereavements consequent upon the prevalence of cholera and its accompaniment, death. Had the members been in an isolated position they could not have done this generous work of charity, and it serves to illustrate the refining and elevating tendency of the principle of social sympathy arising and co-extending with the humanising influence of the new system.

THE EFFICACY OF MUTUAL CO-OPERATION.

The great change which had been effected in the conduct of the peasantry by the action of the principles of justice to all at Ralahine, began to arrest the attention of landed proprietors and social reformers, as well as surprised those who relied exclusively on political action for securing peace and prosperity in Ireland. In passing through Dublin I was honoured by a dinner, where sixty gentlemen assembled, some of whom gave warm expressions of approval of my courage, perseverance, and practical measures —at the time County Clare was in a state of insurrection—measures that had effected such a remarkable improvement, and had been reported by visitors, who had spoken in enthusiastic admiration of our peaceful progress and happy community. It was evident to thoughtful men that people, associated on similar principles to those at Ralahine, and federatively combined, would be more easily and economically governed than the present isolated and opposing elements among individuals. The precarious and unhappy state of a large proportion of the population; their secret hostility against the landowners and the whole order of society; their want of interest in the work they perform; their proneness to premature marriages, ought to lead the legislator, by every means in his power, to give up cultivation by day labourers. How this can be peacefully, effectively, and most successfully accomplished has been clearly demonstrated by the methods which I introduced at Ralahine. There were united to the satisfaction of all concerned, the interests of the landlord, the tenant farmer, and the labourer; and there peace, prosperity, and contentment prevailed. The plan at Ralahine secured to the landlord a certain quantity of the crops raised as rent, the excess being the property of those by whose labour it had been produced. This principle of sharing in the profits was appreciated by the labourers and by the mass of the peasantry throughout the district, and explained in some degree the earnestness and the life-and-death struggles to which reference has been made to gain admission as a member of our association.

CHAPTER XXVII.

THE INFANT-SCHOOL TRAINING.

One of the most interesting departments of our society was that of the Infant-school Training. The advantages of education were not duly appreciated by the peasants; their ignorance and hereditary apathy to the benefits to be derived from culture left them unable to comprehend the issues involved in the cultivation of the intellect, the regulation of the passions, and the exercise and development of the higher moral feelings. In framing the regulations for the government of this department, I arranged that the children, as soon as old enough, should be placed in the care of a teacher under the direction of the president. The children remained during the whole of the day under the superintendence of a young woman, a Catholic, who had been trained as a teacher at the Model School in Dublin.

The children's meals were provided for them, and sent to the schoolroom from the dining-room, and consisted of new milk and vegetables. Those from two years to eight were trained to perform little exercises, calculated to keep them cheerful, healthy, and happy; care being taken that the memory was not overtasked, nor the mental powers loaded with lessons beyond their capacity or years. They were amused, as well as instructed in the qualities and uses of objects around them, and every opportunity seized to inculcate a spirit of love and mutual help in their associations with each other. They were supplied with a great number of oblong pieces of wood, or small bricks, about six inches long, and three inches broad which they were accustomed to arrange as toy houses, and were required to keep them in order when not in use. They were trained to keep step, sing, dance, and amuse themselves in their open playground, which was supplied with suitable gymnastic appliances. No punishment was inflicted beyond being sent home or kept for a time from the school, which was felt as a deprivation of pleasurable amusement. One of the great advantages of the school arose out of our organisation, from the convenience it afforded to the mothers, and the confidence they felt as to the safety of their children during the hours of labour in the dairy. the laundry, or the field. The economy and comfort of the arrangements were soon felt and appreciated. Great attention was paid to the cleanliness and habits of the children, and to the sanitary condition of their surroundings. Little cots were provided for the youngest, when tired or weary.

There is a vast waste of productive labour, to say nothing of irritation, in requiring a healthy woman to devote all her time to the nursing and care of children, the preparation of meals, and the drudgery incident to the management of individual, isolated homes of the working classes, in the houses usually occupied by them, where several are crowded into rooms and are compelled to breathe over and over again the pre-breathed air that acts as a poison. While one woman is occupied in every house or cottage in the kingdom in domestic drudgery, at Ralahine one man and a young woman prepared the food required by the whole of the community. As previously stated, the single men and women took their meals in the dining-hall, and the married members could either dine in their own dwellings, or with the single members if they preferred. Their children, they knew, were under safe and judicious superintendence at school, and required no care till the labour of the day was over, according to the season, when both parents and children enjoyed the change of social intercourse at home, to the evident delight and profit of both. There was no exposure of the children to the pernicious influences of the street, as they were under constant supervision with a view to form good habits and character. To this end the young must be suitably employed, according to their age and capacity, throughout every portion of the day, until their habits become fixed and established.

During my short absence in England in 1833, the infant-school mistress married Mr. Vandeleur's private gardener, and as the members would not accept him by the ballot, she had to leave the society, as was the case during my absence the previous year, when the marriage of a member with a servant in the family of the proprietor led to his retirement, followed by the wake, at which a faction fight arose in consequence, leading to the death of one of the parties, the dismissal of two members, and the suspension of a third. Whether my absence had anything to do with these changes, I cannot say, but probably I might have had some little influence in moderating the results if I had been present. On my return I found the infant school closed, and the mothers kept at home to take charge of their children. Under these circumstances the school was placed, by my advice, under the

SUPERINTENDENCE OF MRS. CRAIG.

A great improvement was soon visible in the progress of the children, and visitors were much gratified with what they saw in the educational department. It was here the advantages of association were most strikingly visible, appreciated, and acknowledged. While the youngest were carefully attended and amused, the elder pupils were trained so as to render the task of instruction a source of pleasure. The acquisition of a knowledge of letters was a source of recreation at short intervals, and pictures of animals and natural objects were made to assist in the knowledge of the meaning of words.

Mr. Pare was more gratified during his visit with what he saw in the training of the children at Ralahine, than any other department, and referred to his recollections some forty years after his visit as a vivid remembrance. He says "The effect of the training on the children was to me—and, in fact, to all who witnessed it—most surprising. They had a fine, healthy, and active appearance; and the pleasing exercises which occupied their attention had a most happy effect on their countenances. The daily ablutions, the regular full meals, although so simple in kind, together with the amusements, contributed wonderfully to improve their manners and expression, which was very superior to others of their class."

"For a short time the school had been left without proper supervision, when Mrs. Craig, at her husband's suggestion, and with the approbation of Mr. Vandeleur, undertook the post of superintendent. Mrs. Craig possessed a powerful and happy influence over the children at Ralahine, and governed without appearing to govern, which is the very *acme* of the art of government. She was fertile in resources for arresting and keeping up their attention, for amusing and instructing them without fatigue or undue excitement, and for cultivating their best affections. She, however, was a Protestant, and owing partly, no doubt, to the attempts at proselytism then making by Protestants in Ireland, a vague apprehension was awakened in the minds of some of the members that instruction would be given to the children contrary to the religious tenets of their parents, all of whom were Catholics. It was a mistaken belief that Mrs. Craig had altered the form of prayer used by her predecessor

In consequence of these rumours Mrs. Craig withdrew from the school, and addressed the following letter to Mr. Vandeleur :—

Dear sir,—As it is probable I shall not have a personal interview with you for some time, I take this mode of making a few statements, which I deem necessary, as I have withdrawn my attention from the infant school, having ascertained that an impression exists in the minds of some of the parents that I am disposed to teach the children doctrines contrary to those professed by themselves. Now, as I do not think that any person has a right to dictate, or enforce their own peculiar religious opinions upon the child of another professing a different creed, it would not be agreeable to my feelings to attend the school whilst the impression above alluded to remains. When I came to Ralahine I found the infant school without any suitable person to attend to the education or morals of those interesting little creatures ; and, at your request, I paid what attention my time would permit ; and so far was I from interfering in this matter that I followed up the same form of religious prayersand thanksgivings which I found had been previously taught by Mrs. O'Dea. After making the above statement, and at the same time observing that I consider the proper education of the children as the most important feature in the society, I hope you will not deem me disobliging when I inform you of my intention to absent myself from the school altogether, unless the parents desire it otherwise ; and, although my services have been gratuitous, yet, I can assure you that I have felt considerable pleasure in doing the little I have for them.—I am, dear sir, yours respectfully,

MARY CRAIG.

This letter, read at a general meeting of the society, had the effect of removing the doubts of the parents, and Mrs. Craig was unanimously recalled to superintend the school. It was the ignorance of some of the parents that led them to mistake the delight of their children in their school, and affection for the governess, as proofs of a sinister and proselyting purpose. It was a misapprehension; but when we take into account the lamentable ignorance of the mothers, and that their children were under the guidance of a stranger and a Protestant, while the enthusiasts were endeavouring to gain converts from the Catholic Church, it need be matter of little surprise. The hatred against the system of tithes, then so harshly enforced in the massacre at Rathcormac, and outrages in other districts added to the bitterness of antipathy which prevailed against the Protestant vicar, and the tithe proctor, supported as they were by the police and the military.

It was a constant remark among visitors that the children were much superior to others of their class, in their frank, open, and intelligent countenances, which beamed with delight, and became even beautiful from the expressive animation with which they were enlivened when under the excitement of their exercises or amusements. They were free from all fear of harsh correction or punishment, and were therefore natural, open, and truthful in all their movements and expressions. The secret of refined culture in expression, so very different between the children of the poor and those of the cultivated classes, lies in the continued manifestation of these happy, inward emotions, which can only be sustained under social arrangements similar to those existing at Rahaline. The rude, angular, and irritating training of the streets—where each is often encouraged to seek his own advantage at the expense of his companions, while the weakest goes to the wall—will account for much of the selfish, sinister, and suspicious expression and bearing of the young, which acts and reacts so as to fix a vulgar aspect on many a youthful countenance.

The proprietor had a family of very fine children, who had the advantages of governesses, and professional teachers, in music, drawing, dancing, &c., and were indulged with many luxuries unknown to the children of Rahaline; yet, he was frequently heard to say that the children of the members—the offspring of peasants—seemed more contented and happier than his own.

RELIGION AND MORALS.

In the formation of the society, provision was made in the rules that perfect freedom should be allowed in the expression of opinion in accordance with regulations twenty-six and twenty-seven, which run thus:—

"That each individual shall enjoy perfect liberty of conscience and freedom in the expression of opinions, and in religious worship."

"That we each observe the utmost kindness, forbearance, and charity for all who may differ from us in opinion."

Admirable as these rules were in spirit and in practice, it was seen that experience taught us the fact that others outside the society were less charitable and forbearing. Mr. Finch, in his account of Ralahine and the truly liberal spirit which prevailed, observes, in his ninth chapter, that:—

The only religion taught by the society was the

unceasing practice of promoting the happiness of every man, woman, and child to the utmost extent in their power, without regard to country, sex, sect, or party. Hence, the Bible was not used as a schoolbook; no sectarian opinions were taught in their schools; no public disputes about religious dogmas, or party political questions took place; nor were members allowed to ridicule or revile each other's religion; nor was there any attempt at proselytism. Perfect freedom in the performance of religious duties and religious exercises was guaranteed to all. The teaching of religion to the children was left to ministers of religion, and to the parents; but no priest or minister received anything from the funds of the society. Nevertheless, both Protestant and Catholic priests were friendly to the system, as soon as they understood it, and one reason was, they found these sober, industrious persons had now a little to give them out of their earnings, whereas formerly they had been beggars.

Sunday was really a day of rest, three members were told off in rotation to attend to the feeding of the cattle and the cooking of meals, and the remainder were at liberty to use their own discretion as to what they did or whither they went. If attendance at chapel was evidence of improvement, the young people, when we had been in operation a few months, were better Catholics than they were before we began. The operations of the society had, in fact, improved the morals of the people, not only at Ralahine, but in the whole district. Outrages ceased, murders were no longer apprehended, landlords returned to their deserted mansions, and although, as we have seen, they had promised to advise the lord lieutenant as to what could be done, they felt secure in their position, and again left the people in their accustomed ignorance, poverty, and wretchedness.

CHAPTER XXVIII.

VISITORS AND THEIR OPINIONS.

During the second year of our association at Ralahine a great change became manifest on the part of many who had prognosticated before we began that it would be impossible to induce the Irish peasantry to dwell harmoniously together, and to give the willing assistance necessary to pay the rent required by the landlord. This adverse opinion seemed justified by the ignorance, superstition, and violence manifested by the discontented labourers. The lawless condition of the south of Ireland, and especially that of county Clare, has already been indicated in the early portion of this history. The harshness of the landlords, the reduction of small holdings, and the increase of grass and grazing where little labour was required, all contributed to develop a state of fear and dread of violence, which made obnoxious landlords seek security in flight from their mansions; and even those who had shown practical sympathy with the people were compelled to relieve the apprehensions of their family by placing them in safety, and their mansions under the care of the armed police force, as was the case at Ralahine.

In the early portions of this history it has been shown with what suspicion I was received by the peasantry; the difficulties that met me in my endeavours to awaken their sympathy and attention to the "new system" proposed for the amelioration of their condition, and to the advantage of both the landlord and the labourer.

After we had been in operation for a few months we attracted the attention of many persons interested in the social progress of the people. It was seen that although the members forming the society were of the average uneducated Irish labourers, with all their prejudices and hatred of the landlord class, they were now, under new circumstances, which made them instruments in their own improvement, quite changed in their thoughts, feelings, and conduct. They were no longer destroyers of property, nor did they join in midnight meetings to decree outrage or murder in revenge for some real or imaginary wrong. Many persons visited the society for the purpose of satisfying their doubts, others came in a spirit of inquiry, and to ascertain how far our plans promised a solution of the difficulties which surrounded the land and labour question. The Government had no remedy beyond coercion; the jealous rivalries of sectarians left the people in their lamentable ignorance, while evictions and emigration were the remedies adopted or recommended by the landholders and their agents. Amidst the conflicting opinions of the various parties offering their counsel as to what was advisable to give peace to the country we steadily pursued our course, and in the second year of our progress we became, as it was said, like a "city set on a hill," attracting the observation of men of thought who had devoted great attention to social and political questions. Among others, we were visited by Mr. Wm. Thompson, a disciple of Jeremy Bentham, and author of a volume on the "Distribution of Wealth," and other works on political and social economy, in which he sustains in an argumentative and exhaustive manner, the principles of mutual co-operation and association for the production, distribution, and consumption of wealth. He was highly gratified with what he saw at Ralahine.

Mr. Thompson contended that in social efforts it was necessary always to keep in view the complicated nature of man, for without this constant reference to it, the regulating principle of utility is sacrificed, and the grand object of political economy, the indefinite increase of wealth or its yearly products, become worthless objects, consigning to the wretchedness of unrequited toil three-fourths or nine-tenths of the human race, that the remaining smaller portion may pine in indolence midst unenjoyed profusion. He contended that it is not the mere possession of wealth, but the right distribution of it, that is important in the community. It is with communities or societies as with individuals. Men cannot be happy without the physical means of enjoyment, which in all civilised societies consist chiefly of wealth. It is not the multitude, but the use and the distribution of the objects of wealth, with which society is chiefly interested. He held that "force, fraud, chance, prescription, are everywhere the main arbiters of distribution, and have almost frightened reason from daring to contemplate the mischief they have made." Mr. Thompson was much pleased with the simplicity and economy of our social arrangements. He was the owner of large estates in county Cork, and told us he should leave his property to establish an association on a system somewhat similiar to what he saw at Ralahine. On leaving us he presented me with a copy of his work on practical instructions in the formation of communities in connection with agriculture and manufactures. This book is a very valuable one, and is full of practical suggestions both in relation to tillage and the architectural arrangement of dwellings of an economical construction. He died in 1833, before he could realise his intentions, but left Mr. Pare and others trustees of his property. The will, however, was opposed by some distant relatives, who contested the legality of a demise of land for the purpose intended, and set up a plea of insanity, which was said to be proved by the nature of the bequest. I had known Mr. Thompson by personal acquaintance with him in England, and among all the men I have since known, I have never met one more under the control of reason and reflection. His physical constitution was not strong, and he maintained a constant control over his appetites, passions, and emotions. His temperament was sanguine-nervous, with little muscular power, and a slender frame, a medium-sized head, and a well-balanced brain. He was a laborious student in sociology. During the last twenty years of his life he avoided intoxicating beverages and animal food, as he could pursue his literary studies much better without them. Like his friend, Jeremy Bentham, he bequeathed his body for dissection for the benefit of humanity. At that time there existed a strong prejudice against the medical profession. The peasantry were opposed to the dissection, and threatened a riot at Clonnkeen, Rosscarberry, where he died.

Mr. Thompson in his day was one of the most earnest advocates of mutual co-operation, and, although he is somewhat prolix and redundant in his great work on the distribution of wealth, he was of a very practical turn of mind, as was evident by the pointed inquiries he made as to our working and social relations. The legal contest respecting his property was sustained in the Irish Probate Court, afterwards in the Court of Chancery, and, ultimately, the decision was in favour of the claimants, and the property became theirs—another illustration of the evils of delay in application of means to legitimate and benevolent purposes.

Another visitor, Robert Owen, of New Lanark, one day took us by surprise, by calling unexpectedly. To no one could our proceedings be more interesting. It was through his advocacy of mutual co-operation and colonisation in Ireland, in 1823, that the attention of Mr. Vandeleur had been drawn to the subject, and led him to adopt measures for ameliorating the condition of the people he employed. Mr. Owen passed many months in investigating the condition of the people—he dwelt in the palaces of the wealthy, and held large meetings in the Rotunda, which led to the establishment of "The Hibernian Philanthropic Society." The Lord Mayor presided, and was supported on the platform by men of great influence—amongst others the Duke of Leinster, Archbishop Murray, Lords Meath and Cloncurry. Mr. Owen submitted a series of calculations showing the abundant means existing in Ireland to place the whole population of that country in a most desirable state of permanent high prosperity. Mr. Owen subscribed £1,000, General Brown £1,000, Lord Cloncurry £500, and other gentlemen £100. Nothing practical resulted from these proceedings, but the proprietor of Ralahine found his efforts to establish a linen factory completely frustrated, as the agricultural labourers had great repugnance to sedentary occupations, and in 1830 and the following year the discontent of the people had broken out in open insurrection, outrage, and murder.

Mr. Vandeleur was strongly impressed with the accuracy and force of the statements published by Mr. Owen, showing what the industry of 1,000 persons could produce if employed on 1,000 acres of land of medium quality of soil, and enjoying the advantages to be obtained in

ASSOCIATED HOMES.

It was known to Mr. Owen that I had been led to the investigation of the aims and benefits to be derived from mutual co-operation through the report of Dr. Macnab on the educational institutions of New Lanark, and it was natural that he should visit us without any preparation to ascertain how far we had attained success, or sustained the accounts he had heard of us in London. We had not, of course, realised the conceptions of the philanthropist, but when he knew the materials with which we had to work, he was evidently very much gratified with our progress. He subsequently, when on a visit to Wisbeach, in consequence of results which had sprung mainly from my exertions in promoting co-operation in Cambridgeshire, wrote in reply to my inquiries as to his impressions of the work done at Ralahine, as follows, with the hope that landlords, capitalists, and farmers, would obtain the services of Mrs. Craig and myself, to put the plan in operation in England. "It appeared to me," wrote Mr. Owen, "from a full inspection of your proceedings at Ralahine, that, considering the means which your association possessed, your arrangements to produce and distribute wealth, to educate and form the character of the people, and to govern them, were excellent and carried on in the true spirit of co-operation. The people appeared to me more happy than any of the same class in any part of Ireland, which at various times I have visited, and that the proprietor was not only in much greater security from injury of person and property, but he expressed to me in very strong terms the great satisfaction which he daily experienced from being with the co-operators, and witnessing the extraordinarily successful progress which you were making towards independence for yourselves and for him."

I have already given extracts from Mr. Finch's book, the result of his visit, which extended over three days, and as I appointed one of the labourers (a member of the committee) to accompany him and answer any of his inquiries, he had the best possible opportunity of ascertaining the working operations of the society. He was a man of thorough business habits, a close, accurate, and critical observer. He was an earnest advocate of the temperance movement, and of mutual co-operation. As a merchant he had repeatedly traversed the three kingdoms, and was well aware of the condition of the people. In closing his series of fourteen letters on Ralahine, he says—"When I have rested and have more leisure I promise the second part. In the meantime I call upon those parties who know most about the institution at Ralahine to confirm my account, supply my deficiencies, or point out my errors: I shall be obliged for either. I seek truth, and endeavour on all occasions to write and speak the truth, without mystery or fear of man. I have no interest in error, and never speak truth from selfish, mercenary, or malignant motives. I love every human being, and wish to see every human being wise, virtuous, and happy."

He did not live to realise his wish to complete his work, and the appeal he makes, he repeated to me in private, urging me to give the history which is now approaching its conclusion. We have seen that Mr.

Finch offered to lay his report of Ralahine before a Select Committee of the House of Commons, to show that the new system was important to landlords. He held it impossible to devise arrangements more easy, practicable, and economical than those he saw at Ralahine. He viewed our poor laws, charity schools, charitable institutions, and even our national colleges as paltry, mean, and despicable when compared to the simple, rational, and natural arrangements and social advantages existing at Ralahine.

In the autumn of 1833 we were visited by Mr. W. Pare, who was on his way to Roscarberry, county Cork, to take possession of the estates of Mr. Thompson, who had left them for the establishment of similar institutions to that he had seen at Ralahine. Mr. Pare stayed with us a day and a half, and was very strongly impressed with our educational influence. He has recorded his impressions in a volume published under his name, and to which I largely contributed. In the closing chapter he appeals to the patriot, to the philanthropist, and those who seek chiefly their own interest, and promises "for each of these aspirants for fame, for worthiness, or for wealth, a way to gratify to the fullest their desires by following the example set at Ralahine. This done, the patriot would see the condition of his country would rapidly change for the better. Order will rise out of chaos; kindliness, generosity, and mutual assistance will replace revenge, rapine, and murder; and, instead of a herd of discontented and rebellious helots, he will witness a nation of loyal and happy freemen.

"The philanthropist will see in motion a machinery which will quickly become self-acting and self-sustaining, by means of which he will witness the gradual rise of the labourer and his family from ignorance, poverty, intemperance, and brutality, to intelligence, competence, sobriety, and comparative refinement.

"The self-seeker will find this the shortest and surest way to secure and increase his income, and at the same time rid himself of his now hourly care, anxiety, and insecurity of property and life.

"And, fortunately, the patriot, the philanthropist, and the self-seeker, each in pursuing this path for his own special ends, must of necessity aid the others in attaining theirs. Though each start with a different purpose, all will arrive at the same glorious result—the emancipation of their common country from poverty, misery, and crime.

"And this blessed result is to be achieved by the simplest possible means, to apply and continue which requires only ordinary business talent, combined with discretion, firmness, and perseverance."

The attention of the Irish Government had been drawn to the success of our proceedings, and it was believed that an intention existed to depute a competent person to visit us and report. This purpose was frustrated by the sudden and unexpected termination to our proceedings.

The visit and testimony of the Rev. Francis Trench, brother of the present Archbishop of Dublin, has already been recorded.

Among many other Irish landlords who visited Ralahine was Lord Wallscourt, who had estates in county Galway, and he was so much gratified with what he saw, that he proposed to bring his farm bailiff, to whom I supplied a copy of our agreement and regulations, with such explanations as seemed necessary. His lordship laid down 100 acres to which he adapted some portion of our plans, and with marked success. Some years after, when speaking of the system in a letter, he said :—

"It answers much beyond my hopes, inasmuch as it completely identifies the workmen with the success of the farm, besides giving me full liberty to travel on the continent for a year at a time ; and upon my return I have always found that the farm had prospered more than when I was present."

This identification of the workman with the success of the farm, was one of the great and grand moving principles at Ralahine, and contributed largely to our success.

CHAPTER XXIX.

HARVEST HOME REJOICINGS.

The progress of our Association, as we have seen, had become very conspicuous, had exercised a marked and acknowledged influence on the moral improvement of the people; the rarity of agrarian outrages, the increased confidence of the landlords, shown by their return with their families to their mansions, together with the satisfaction at the very successful harvesting of all the crops, it became an expressed desire of the members, embodied in a determination of the committee, to hold a Harvest Home rejoicing as an expression of thankfulness at the harmonious results of past exertions. One of the most striking proofs of the power of the system of sharing in the profits of results, in awakening the impelling motives to exertion, was seen in the fact that at harvest time the members of the society would voluntarily work longer than the time specified; and I have seen the whole body occasionally, at these seasons, work with an energy, and accomplish such great results by their united exertions, that each and all seemed as if animated by a wild and enthusiastic determination to achieve some glorious enterprise, and that, too, without any additional stimulant being administered to them in the shape of extra reward, beyond what might be realised by their share of the profits after the rent and interest of capital were discharged.

Although the members were only paid the ordinary low wages of the district, yet the advantages derived from the economy of associated efforts were so distinctly manifest in their social arrangements in relation to food, dwellings, and the training of their children, that there could be little doubt as to the influence of these advantages enlisting the zeal of the members in manifesting the industry, care, and thrift so unusual in the labourer employed for the sole benefit of another. A man working without any chance of gaining by his extra labour is deprived of the spirit stirring energy essential to rouse the latent forces of the brain that nerve the physical powers to their highest and best expression. It has been well said by Sedley Taylor, of Trinity College, Cambridge, that "an agricultural labourer in his employer's field, and the same man in his own garden, present two surprisingly different standards of activity. The reason is obvious. In the former case there is the deadening certainty that additional effort will receive no additional wage; in the latter, there is the enlivening hope that every stroke of more efficient labour will meet with its due reward. Accordingly, a medium standard of work is all that can be reckoned on from men labouring for fixed wages, and even to exact this requires constant and vigilant superintendence. If, on the other hand, some stimulus, similar in its effects to that of ownership, can be brought to bear on men industriously employed, there is reason to expect that their standard of work will be correspondingly raised.

"The prospect of sharing in profits is manifestly just such stimulus, and, therefore, enhanced efficiency of labour may legitimately be anticipated from its operation. Improved work spontaneously given brings with it, in general, increased production, better quality, and a diminution in the costs of superintendence. It, therefore, means in many, perhaps in almost all industries, enhanced profits. Here, then, is the economic basis on which participation rests—the creation, by the more efficient labour called forth under its influence, of new profits which do not accrue under remuneration by fixed wages only."

It may be easily conceived that in the early days of our experiment we were not without anxiety, in consequence of the opposition, the suspicion and moody feelings of the people. It was not till after the ballot of the whole of the members had been taken, numbering forty adults with twelve children, that the members found that the committee whom they had elected had really the sole power of appointing everyone to their allotted duties, and without the hated arrogance of a detested steward, that the change became manifest in the intelligent eagerness to discharge their duties. The industry and skill of hitherto apathetic, sluggish, and sullenly silent labourers became very manifest, affording a gratifying testimony to the beneficial influence of the plan which awakened the spirit of the workman by giving him the impelling motive for exertion, by assuring him a share—although but a small one—in the profits arising out of his increased exertions. From the moment this confidence was obtained, the affairs of the community worked harmoniously, except in the incidents recorded. Two members had withdrawn from the society as there was no suitable occupation for them, two were dismissed through a quarrel outside the society, and two retired, as they acted contrary to the law relating to premature marriages. The comfort, real enjoyment, perfect freedom and individual independence of every member became a subject of honest pride and boast of the humblest amongst them, and a matter of wonderment, not only to the entire county, but to the whole of that part of Ireland.

We had numerous applications during the spring and summer of the second year of our operations. The social privileges of the members were found to be far greater than those of small farmers in the neighbourhood, living in miserable cabins or isolated dwellings, surrounded with all the discomforts inevitable in the habitations of the Irish peasant.

The number of members in the autumn of the second year was as follows :—

Adult men	35
Adult women	23*
Orphan youths under seventeen years ...	7
Children under nine years and infants ...	16
Total population	81

An abstract from the daily labour sheet in April of the above year will indicate the several occupations of the adults during one day. As the children and youths under seventeen years of age did not draw anything in the shape of wages, their labour was not entered :—

Fifteen men employed in spade cultivation.
Four men making up compost.
Four men carting manure.
Four men at four ploughs.
Three attending milch cows and cattle.
One steaming potatoes for meals.
One at butcher's work.
Three at carpenter's work.
Two at smith's work.
One storekeeper.
One secretary.
Eight women at agricultural operations.
Three women at dairy and poultry.
One at domestic arrangements.
One mistress of infant school.

* Two of these did not receive wages.

Owing to the great success which had resulted in the zealous and harmonious co-operation of the whole of the members, old and young, in gathering in the crops, it was proposed to decorate the last load of wheat with garlands and flowers on the field where it hăd grown, before it was drawn home to the rickyard. This favoured load was from the twenty-acre field, which in the spring of the previous year had been brought by spade cultivation from the barren condition under which it had remained from all time to that of a fruitful field of potatoes, followed by the best crop of wheat on the farm, and which formed a matter of surprise to the surrounding district at the rich return realised by co-operative labour. The heavy crops were, doubtless, attributable to the deep spade tillage and the ample supply of complete manure. The members rejoiced at the acquisition of twenty acres of productive land, which increased their prospective share in the profits, without any additional charge for rent. They knew, and felt that their reward was in proportion to their care, industry, and devotion to the common interests of the society. It was determined, with the approval of the president, to mark the occasion with a procession.

THE FIRST AND LAST FESTIVAL.

The last load to the stackyard was heralded by music. A procession of the whole of the members with the youths and children, headed by the secretary, mounted on a grey pony and wearing a silk sash, on which was printed in letters of gold, "Each for All," marched round the boundaries of the estate, to the surprise and delight of the people in the neighbourhood, who, in many instances formed a marked contrast to our members, some of whom, like them, were, on their admission, clad in rags,

—and in one family the children shoeless and shirtless —were now in the proud possession of two suits of clothes in addition to savings varying from five shillings to five pounds in stored-up labour notes.

When the procession reached the front of the Hall, the residence of Mr. Vandeleur, the landlord and president of the society, that gentleman came to the door of his mansion, and in the presence of his family and several visitors, addressed the members in a very suggestive speech.

He expressed the great satisfaction he felt at the progress the society had made for the time it had been in existence; at the harmony which prevailed in the social arrangements of the people; their evident comfort, prosperity and contentment; contrasting the present happy state of Ralahine, and the quiet condition of the county with what it was when his family were compelled, from the dread of outrages and murders, to leave their home in the care of an armed police force. He congratulated them on the operation of the new system which had accomplished a success greater than he had expected, and hoped that other landlords would appreciate the advantages of giving those they employed a share in the profits realised by mutual co-operation, as might be seen in the evidence given by the large crops raised on the hitherto waste land, made richly productive and profitable in a large crop of potatoes during the first year, and followed by the splendid crop of wheat, the last load of which they were now cheerfully leading home. He also expressed his satisfaction at the methods adopted by the committee in the management of the farm, but regretted that measures had not been adopted to utilise the water-fall—twenty horse power—then running useless in the water-course in Dereen, not far from the new wheat field, for the manufacture of frieze cloth, which he considered likely to prove very remunerative to the society.*

But the most gratifying results of the association, he held to be the happy effect of their social arrangements, and especially the training of the children and youths to habits of industry and intelligence in the use of their resources, for their own individual improvement and happiness.

If the society had been in possession of capital, the manufacture of freize cloth or linen might have been successfully carried on. As Mr. Finch had said, no manufacturer could possibly compete with the society, in the possession of such water power all the year round. But it would have been premature to attempt manufactures at that early stage. There were, however, some departments which might have been encouraged and developed. The poultry yard might have been extended, and as we had about twenty-five acres of clover, a large amount of honey might have been raised by a stock of bees.

This address of gratulation from the president and landlord for the past success, and encouragement for the future, was received with hearty cheers, and the procession marched forward to the lecture hall, where they partook of refreshment; a marked feature of which was that the bread made under the direction of Mrs. Craig, and liberally supplied, had been, for the first time at Ralahine, made from the new wheat grown on the new land,

* In the Agreement made between the trustees and the landlord, it was stipulated that "if the water power in Dublin was not converted to some use before twelve months, that it ceases to be included in the property handed over to the society."

threshed, garnered, and ground on the estate. To the members the art of making and baking their own bread was then a novelty and a mystery. Hitherto the potato had been their chief stay, and would have proved utterly unequal to the physical requirements of labour if they had not had the essential support derived from the daily supply at every meal of a pint of new milk.

The evening of that day was spent with a happy hilarity, a merry festival of dance and song. Attempts were made to introduce the English country dance, but the Irish jig maintained its supremacy among the adults. It was an easier task to train the children to the mazes of the quadrille, and the exercise had a salutary influence in training them to order, method, discipline, and courtesy. The people were highly gratified with the proceedings of the day, and passed a vote of thanks to Mrs. Craig for the time and attention she had bestowed in decorating the rooms, and for the lesson she had given in bread making, an art, it was hoped, their increasing prosperity would enable them to sustain.

CHAPTER XXX.

SIGNS OF SUCCESS AND SATISFACTION.

The New System had fully established itself, and proved its adaptability to our social requirements, for the society was now in full career of success. It is true we had not realised all the benefits which we anticipated would arise from increased experience and a more widely diffused comprehension of our principles of action; but there was an admitted increase of security, peace, and happiness arising from the associated action of our community.

Before I had drawn up our plan of organisation, or undertaken the trusteeship, it was the apparent interest of the labourers to destroy the property, the peace, and in several instances the lives of men among all classes, the landlords, the stewards, or obnoxious witnesses; and no advice nor force could divert them from their violent course of organised opposition to the landlords, and combined resistance to their requirements; but now that their circumstances and their real interests were in harmony with the prosperity of the landlord and everyone else, they acted in unison with the interests of all around them. Their minds, or powers of comparison and judgment, were quite apprehensive enough to lead them to a due appreciation of their altered conditions, and to teach them the consequences of their own actions, as well as prompt them to proper conduct. Surrounded by favourable circumstances, they were necessarily virtuous and contented. Formerly, the labourers were in want, and necessarily discontented ; it is a difficult problem to inculcate quietness and placidity to an empty stomach, and hence they were turbulent and unreasonable ; but now, on the same estate, they were industrious, peaceful, and hilarious. Their nominal wages were the same, but their social arrangements now gave them immense advantages, which they knew and acknowledged. Formerly they were never sure of the future ; a petulant impulse, or wayward passion, on the part of the steward—as had often been the case —might take away their means of subsistence ; now the future of employment was under their own control. Until three years previous, they were at the beck and call of a harsh taskmaster, and, however much they exerted themselves, they got no thanks, many curses, and much irritating villification. Now, no man was master, all were equal, and none provoked vindictive utterances. However valuable might be the labourers' experience, and promising their power of suggestive advice, none dared to admonish their master. All were silent, sullen, and inwardly in opposition, and often outwardly contentious in action, but now the "Suggestion Book" bore many proofs of practical thoughtfulness, from men, hitherto despised, because they were labourers, and withal, very poor and needy.

It is the custom of men learned in the languages of schools, wise in the wisdom of political economy, and experienced in government, to condemn those who are idle from compulsion, and to censure those who have nothing left to preserve, for their lack of providence and thrift. They also rail at the discontented utterances of men of intelligence, who from sympathy, feel the inexorable iron of wrong entering their hot life blood, and tearing out their heartstrings; and who, while admitting that "Force is no remedy," yet apply "Coercion" to compel acceptance of methods that are, at best, mere palliatives, not anodynes or effective social civilisers.

These adverse conditions arise out of the existing antagonising relations between capital and labour, or the employer and the employed. In the Middle Ages war was almost the only business of life—every castle was a fortress—the jerkin was thown aside for the coat of mail and the gauntlet. To-day the same faculties of the mind are devoted to business. Men now fight, not with battle-axes or spears, but with bullion, the pen, and the banker's cash book. They do not now enforce slavery on those they conquer, but competition vanquishes labour till it becomes the slave of capital accumulated out of profits, or the results of productive industry. Trade, as now carried on, is a warfare of personal combat, a race of selfishness, wherein "the devil takes the hindmost." The interest of the producer and the distributor prompts both to cheat the consumer, for adulteration is but a "phase of competition." Between the two the labourer is crucified, as he is only allowed to enjoy a fraction of the fruits of civilisation. Genius, with its usual bounteous prodigality, has poured out its riches from Arachnoid's web, but the gems of art, with few exceptions, are secreted in exclusive halls, mainly accessible to the capitalist class, while the sons of toil are restricted to their naked inartistic surroundings. The capitalist dazzles the world with his profuse expenditure, and masses of accumulation and luxury, while the privations of the real producers are disregarded, or little thought about, beyond the standing policy of the commercial and trading system, to keep down the wages of labour so as to be able to compete victoriously over foes at home and abroad, not always by fair fight or competition, but by a system of forced abstraction which the producer yields in low wages, and in taxes so far as he is a consumer.

Art, literature, architecture, and science have been cultivated mainly for the advantage and enjoyment of non-producers. Admitting the influence of heritage, and the varying aptitudes transmitted to "the fittest to live," still, all civilised races of men have the same number of faculties, or cerebral functions, which, if cultivated, would enjoy the brilliant products of inventive genius. Talent and genius are not a class or caste. Cimabue discovered Giotto by observing him drawing the sheep he was herding. Presidents Lincoln and Garfield, as well as Stephenson and Burns, were sons of toil, and could enjoy the true and beautiful.

The Greeks regarded beauty as another name for goodness or excellence. So great was their admiration for their masterpieces of art that they refused to part with a single statue, even when promised the removal of their debts. When Cicero was in Greece he was everywhere reminded of her great men by their monuments. No education exercises so refining an influence over the young as artistic beauty. It has been well said that the ideal of life, as well as in art, has been the blending of the beautiful and tender with the strong and the intellectual, which can only be realised for the producer by the conditions of harmonious, social associations. The youth of Greece would stand enraptured at the statues of Praxitiles. When Cicero was at Salamis he was reminded of Themistocles, and thought of the divine ethics of Plato. The Parthenon told him of Pericles, of Marathon, of Miltiades and his ten thousand. It was the tastes of the people that raised Phidias to grace the frieze of the Parthenon, where the sculptor has idealised beauty without departing from nature in those marvellous, and all but living, groups of the artist. It was the vivid enthusiasm for the beautiful in sculpture and literature that made the Greeks the great educators in art and literature of succeeding generations. But where are the temples, palaces, and statues to awaken the tender sympathies of the sons of toil of to-day? Why should not the builders of palaces dwell in them? Why cannot the joiner, the mason, and other artisans dwell in mansions? The answer is given by Ralahine. It is simply a matter of association, labour, culture, time, and capital.

Architecture is made to minister to the enjoyment and pride of the capitalist of to-day, while the cabin and the cottage remain in their pestiferous ugliness. It is only a few years since when prejudice held the drag-chain on the wheels of educational progress. It was once held that servants, if educated, would be above serving "their betters." The sons and daughters of service part with their liberty, but not with nature in their contracts; they are flesh and blood, have feelings and emotions, longing for exercise and gratification on their legitimate objects, even in the house of bondage, as well as their masters. But how are they to gratify these aspirations, if they are shorn of all true freedom. The command is everywhere "Work, work, strain every nerve to increase production for our profit, but seek not to know the true 'resources of civilisation,' in the enjoyment of art, literature, or science, and hope not to share in the direction of the work, or to participate in the surplus profits arising out of our mutual exertions."

Science has already added much to the glory, convenience and advantage of commerce, but it has hitherto done little for the social happiness of mankind. It has, however, prepared the way for mental freedom, by breaking the chains and destroying the despotisms that shackled thought and terrified reason. Knowledge and truth were once confined to the few; now they are the birthright of the many, with the recognition of the right of private judgment, and a more universally diffused exercise of the brain. The ultimate result must be the adoption of principles which our association at Ralahine has demonstrated to be true, in proving the powerful influence which lies within the moral force and social sympathy of mutual co-operation. These principles will exercise an ever-increasing power over the intelligent minds of future generations, and especially in the New World.

The general tendency of existing conditions in competitive society is to accumulate wealth and enrich a few, while the mass of the producers must become poorer, and still more hopeless, to throw back the middle-class on the profits derived from labour, till the wealth-seekers, bankers, and speculators upon interest accumulate masses of capital, that can at all times control the products of the people's toil.

Producers, when they become fully alive to the importance of perfect freedom and the power of capital, will adopt, with a resolute purpose, measures for securing all profits on their own labour, while they absorb all rents, interest, and dividends on trading transactions. Rent is a tax upon Toil, and the middleman often a burden too grievous to bear.

This line of associated social action implies a cultivated intelligence in relation to social science, under the control of the enlightened social direction of duly qualified guides. Our members had not enjoyed all the advantages that our plan offered to them, owing to a lack of cultivated intelligence; but it has been clearly demonstrated that their perfect freedom from repressing serfdom, their share in the direction of their own labour and management of the establishment, their social privileges which can only exist and manifest their influence under similar social organisations, had developed a spirit of manhood, independence, and energy, which were not previously supposed to exist. In fact, the labourer before I had drawn up our regulations, and the same man now, in the enjoyment of our system, were two very different individuals in character. Before we commenced he was surrounded by conditions calculated to give activity to his selfishness, his cunning, and his destructiveness. Society was opposed to him, and he resisted. Every other man's interest seemed adverse to his own, and he was against every other competitor. But now he was under circumstances where his appetites were reasonably, though moderately, satisfied; his individual welfare depended on the general welfare; his happiness was coincident with the happiness of all around him. His confidence in the future was assured, and as we have seen, Nature had given us reason for holding a happy harvest home, and the members danced, sung, and leaped for joy at our progress and success.

The members were full of satisfaction with the present, and hopeful as to the future. The harvest was a splendid one. The new land of twenty acres had yielded an ample return for their extra labour, which had been the means of adding to our tillage land without increasing the rent. Six new dwellings had been erected by their own labour. It was also expected that the anticipated surplus would add twopence a day to the wages of labourers receiving eightpence, and one penny to women's wages, being an addition of one-fourth in one case, and one-fifth in the other. These advantages were in addition to those secured by wages, such as a second suit of Sunday clothes, while their children were clothed, fed, and well educated out of the common fund of the society, and all of them had labour notes in reserve.

Beyond these acquisitions and advantages, the association had, by their combined labour and care, produced and delivered to the landlord as rent and interest, for the land, stock, buildings, and machinery, the following quantities of produce:—

6,400	Stones	of	Wheat.
3,840	,,	,,	Barley.
480	,,	,,	Oats.
10	Cwts.	of	Butter.
30	,,	,,	Pork.
70	,,	,,	Beef.

These great results had been realised within three years at Ralahine, and others, with the right men, might follow our example. Leaders and organisers, sufficiently enlightened as to the principles involved in the new system, or science of society, with all the higher "resources of civilisation," could call into existence similar associations in a short time, and establish them in every county in Ireland; and what a wondrous change would be seen in the green isle of the ocean! As it has been truly said, if our system had been allowed to continue, its example might have helped to make Ireland a paradise of peace.

CHAPTER XXXI.

THE LAST ACT.—FLIGHT OF MR. VANDELEUR.

Hopeful, however, as we had reason to be, and sanguine as we all were in expectation of a glorious morrow and a happy future, our day-dreams were doomed to a sad and sudden awakening to sorrow, disappointment, and grief by the disappearance of our president, followed by newspaper announcements, in head lines with bold Roman characters—

"FLIGHT OF JOHN SCOTT VANDELEUR!"

Through a deplorable weakness to which his mental constitution—a family feature—laid him open, he was often attracted to his club in Dublin, where he associated with men of the highest rank in the country; and it was said he had indulged in the vice of gambling, which brought "debts of honour" he could not meet. His imprudence involved all he possessed. There must have appeared to him no hope of extrication; and to avoid the shame, opprobrium, and disgrace which awaited him on all hands, those nearest and dearest to him not excepted, he resolved on instant flight.

The intelligence came upon our happy community like an unexpected thunderbolt. At first the members were stunned with the news, the effects of which they could hardly realise. The way in which the information was received was painful, and extremely distressing. Upon its confirmation, I heard women and even stout men grieve piteously. These children of feeling and sympathy embodied their sorrow in the manner usual at the bier and the grave of those most dear to them when suddenly snatched away by the hand of death. Without education and without study, their lamentations over the dead were often conveyed in such pithy, forcible, and audible language, whether in praise of the departed, or grief at his loss, that the utterances were often very touching and suggestive. I once attended, at a little distance from Ralahine, the funeral of a respectable farmer, accompanied by numerous friends of the deceased, where the custom of Caoinan —probably of Eastern origin—-was practised. Formerly, until Elizabeth's reign, the bards prepared the Caoinan or wail for the dead. The women wore mantles or cloaks of the Spanish fashion, but no head dress, and many of them had dark eyes, dark hair, and a stately step, bespeaking Milesian descent. The mourners arranged themselves in procession, and as they passed along the hill side, they occasionally burst out into a melancholy wail or song, which I did not understand, but it had a weird-like effect as it undulated along the crisp, sharp air of the hills.

This sympathetic audible utterance, less practi ed now, prevailed at Ralahine among the peasantry. As the house occupied by myself and Mrs. Craig was near the cottages of the married members, the wailing of the people in the night had a sad, mournful, and heart-rending effect. It requires a knowledge of the habits to which I refer to fully appreciate the depth and force of their audible sorrow when intensely moved. It was afflicting in the extreme to hear in the still hours of midnight, the wild wail, as if over the dead. Whether it was owing to the painful shock I had received in the frustration of all my reasonable hopes, and the apparent failure of all my labours which had, in fact, been most successful, and satisfactory to all concerned, and would ultimately, if continued, be likely to affect all Ireland, coupled also with an intense anxiety that the Labour Notes accumulated as savings in the hands of the members should be paid, or cashed and cancelled at once, I know not, but I was much agitated and agonised, when, at intervals, the people would cry aloud from the depth of their sorrow, "Ohone! ohone! SHAHN VANDELEUR, why did you go from us? Ohone! VANDELEUR! why did you leave us? Why have you left your own Rahlaheen?"

The sudden disappearance of the President, the disappointment, and the distress of mind, had a very serious effect on my health, and, although at the time only twenty-nine years of age, I arose next morning with many grey hairs, caused, doubtless, by the intense action suddenly thrown upon the nervous system. I might say with the prisoner of Chillon, if not exactly in the words of Byron—

My hair was white, but not with years;
It grew to white
In a single night,
As men's have grown from sudden fears.

As the President had called on me two or three times just before his departure in his family carriage, it was assumed that I was aware of his intentions, and had a knowledge of his destination. Under this impression, Mrs. Vandeleur sent a messenger, requesting me to visit her at the hall. On taking the shortest way, I had to pass through the grove which sheltered a portion of the mansion—the spot where I had been struck on the back of the head with a stone before the society was organised, or the plans proposed to be introduced known and understood, and I naturally thought of the great improvement effected among the people of Ralahine, the whole county of Clare, and the south-west.

I found Mrs. Vandeleur accompanied by the mother of the President, a fine, tall, and venerable lady, bearing the signs of heartfelt grief and distress. Her daughter-in-law was a very handsome woman, the mother of five beautiful children—three girls and two boys. The youngest, Arthur, a fine, lively boy, three years of age, was present. His mother was not tall but rather inclined to *embonpoint*, with a fair complexion, a flushed cheek, unusually prominent eyes, giving a sensuous rather than intellectual expression, and a square forehead harmonising with her earnest piety. Mr. Vandeleur, I heard, received over twenty thousand pounds with Miss Molony on her marriage.

Mrs. Vandeleur said, "As we hear that Mr. Vandeleur called on you, Mr. Craig, just before he left Ralahine, we suppose he told you whither he was going, and where he might be found ?"

I told her that " although her husband had visited me, he had given me no information whatever as to his movements, nor did I know anything as to his address. He appeared somewhat strange, distracted, and unusually reserved. He brought to me a number of papers, and among others the receipt for the money paid for the air pump which you have seen me use in the lecture-room. He also gave me his double-barreled gun."

They could scarcely credit me with speaking the truth, but I repeated my testimony, and assured them that I was as much astonished at his departure as they were, and I found myself in a very perplexing situation. Through the advice of Mr. Vandeleur, I had been induced to bring Mrs. Craig from England.

I was now thrown into a false position with regard to the wages and the savings of the members. It seemed to dawn upon the ladies that I might be telling the truth, and this only appeared to increase still more the agony already too great to be borne. The tears rolled down the cheeks of these two suffering women, whose sorrow I would have removed if I possibly could.

My own feelings were much disturbed—the heart was too full. I felt as weak as the child. A gush of tears gave relief, as I left this house of woe. If the poor have their troubles, the wealthy are not without their sorrows.

The recollection of these scenes, although whole oceans of events have intervened, makes the heart sad, and the feelings are awakened by the echoes of the past that now fall on the ear, and cause tears to flow and wet the paper on which I write. I would they could wash out the record of the present chapter, then there would be

SURCEASE OF SORROW.

But Mrs. Vandeleur, weighed down by her misfortunes, in having to watch the progress of disease in her three daughters, who all died of consumption, while her eldest son met a watery grave in bathing, fell into a decline and died broken-hearted before ten years had gone, when she passed away, leaving her only son, Arthur, the last scion of his house and family branch.

Having alluded to a family feature in the Vandeleurs, I may remark that special tendencies have much to do with particular habits. A genius for music, mechanism, or mathematics will be the result of family aptitudes, as in the Brunels, the Gregorys, Napiers, &c. It is so with hopeful speculators. Colonel Vandeleur, the father of J. S. Vandeleur, was fond of racehorses. He laid down the headlands of Granahan, our largest field of seventy acres, as a racecourse and exercise ground for his stud. He must have had a large development of Hope. His son—our president—had large hopefulness. His wife had large Faith, as indicated by her square forehead, her strong religious feelings, and conduct. Arthur Vandeleur had, as shown by his conduct, the Hope of his father and the Faith of his mother, as illustrated by his life and his portrait, given with his interesting memoirs, which represents him as a major in the Artillery with a medal and clasps on his breasts, for service in the Crimea. The portrait has a strong resemblance to the handsome face of his father, and the full dark blue eyes and square forehead, and Wonder of his mother.

The life story of Arthur Vandeleur is a striking illustration of the evils of military training in schools. He was sent to Mr. Hare's school, at Delgany, county Wicklow. The school was attended twice a week, for his own amusement, by a retired captain of a Highland regiment, and in uniform, who made "quite a disciplined corps in company drill" among the pupils.

With his mother had gone the solitary influence which might have led him to select a more useful mode of life than the profession of slaughter, for, as Dr. Proteus says,

> One murder makes a villian,
> Millions a hero.

A nomination was obtained for him by Lord FitzGerald, and the barracks and the army secured what nature meant for the platform, the pulpit, and the professor's classroom. He went to Woolwich, passed his examination, and in 1847 joined his regiment at the Royal Artillery

Barracks, Woolwich Common. He soon found how difficult it was to sustain that incongruous character —the successful soldier and the pious Christian. Unless in national defence of home and duty, I look upon all soldiers as unnecessary

Murderers riding deathward on Slaughter's Red Car. They are non-producers and destroyers of the fruits of toil, and of life.

When the Crimean War required the flower of the English army, Arthur Vandeleur and hundreds of young Englishmen fought the useless fight. Arthur was attached to Maude's battery. He rode beside the foremost gun up the slope at the Alma. The Russians, under their system of large masses of men in close order, moved over the plain opposite to the English left. It fell to Arthur's hand to fire the first shot at the foe. The distance was soon found. The ball was seen to cut a lane through the heavy Russian ranks. Other shots, and then the Russians fell into confusion. The enemy's guns were of large calibre, and reached a distance of 3,000 yards, and a 12lb. shot killed a shaft horse as Vandeleur rode beside it. His battery got across the river Alma, enfiladed the enemy's entrenchments, and they put every Russian to the sword. And there, as "Childe Harold" says :—

There shall they rot—Ambitions' honour'd fools ;
Yes, Honour decks the turf that wraps their clay !
Vain Sophistry ! in these behold the tools,
The broken tools, that tyrants cast away
By myriads, when they dare to pave their way
With human hearts—to what ?—a dream alone.
Can despots compass aught that hails their sway ?
Or call with truth one span of earth their own,
Save that wherein at last they crumble bone by bone ?

When peace was declared, he returned, and married his cousin, Miss Mary Molony, of Kiltanon. He mortgaged his estate of Ralahine for a large sum, and, although not compelled by law, paid all the debts of his father, an honourable transaction, worthy the good name of the Vandeleurs.

He sustained a school at Ralahine for five years, and paid the expenses of a first-class teacher, but the priests would not let the children attend the school because the Bible was used as a school book. It was well meant on his part, but an error in practice, which we avoided, and there are better class books than the Bible for instruction.

The author of the life of Arthur Vandeleur, in dwelling on his efforts to promote the happiness of others, has the following appreciative remarks:—

Perhaps few people have a better right to make a little dissertation on the happiness of giving than Arthur Vandeleur had. He was always on the alert for an excuse for making little gifts to his friends; gifts which seemed to be chosen with a sort of intuition as to that which would afford most pleasure. Birthday, Christmas Day, New Year's Day, Midsummer Day, days of parting, and days of meeting, all alike were pressed into this service. And to the needy and destitute, his charity was only bounded by the righteous dread of debt, and desire neither to spend nor to give a single farthing which did not positively belong to him; in order that, like Longfellow's blacksmith, he might—

> Look the whole world in the face,
> For he owes not any man.

All who appreciate the great and grand principles of mutual co-operation should cherish the memory of the founders, and when they have their children with them at Woolwich they should show them the monument "raised to the memory of Arthur Vandeleur, of Ralahine, county Clare, who died at Woolwich," and tell them the story of his father, who enabled the writer to show how easily the new system could bring peace, order, prosperity, and contentment to all dwelling on the now discontented Green Isle of the West.

Were I not restricted to the boundaries of a sick room, I would seek out the spot where Arthur lies, and lay a wreath upon his tomb. Co-operators should induce their children to drop a flower on his grave—even if only a wee, modest daisy, it would be as appropriate as the fairest camellia or the richest stephanotis—keeping alive the cause in their memory, and prompting them to carry onwards the blessed principles of Social Co-operation.

CHAPTER XXXII.

FIAT OF BANKRUPTCY AND LEGAL DIFFICULTIES.

If I had reason to be gratified at the success of my efforts at social organisation, and proud of the great improvement that had been realised among the people of Ralahine, where order and harmony had been evolved out of discord, passion, violent agrarian outrages, murder, and social chaos, I had great cause to be disappointed and pained at the issue. Prosperity and contentment had been made to supersede the deepest distress, destitution, neglect, cruelty, and oppression, which had prevailed. Where wrong and injustice had aroused intense indignation, violence, and vindictive revenge, there peace, forbearance, good will, and charity had been awakened. Now a great change had come over the scene. The happy community was shattered, and my position extremely uncertain, and the probable future, dark, gloomy, and hopeless. The past became as a dream, the present a hard and harrowing reality, with a mind anxious and troubled, aggravated by great weakness from loss of health.

The farm had been considerably improved and extended by the mutual co-operation of the members. It was in good heart for succeeding crops, giving promise of future prosperity. The buildings had also been extended by the erection of six new dwellings by the extra labour of the members; while the people themselves had apparently been regenerated by the spirit of the new system of dealing with the land, the landlord, and the labourer; a great socialising improvement had been effected in the educational training of the young, acting and reacting on their parents and friends. All this availed as nothing. Social Co-operation at Ralahine was a thing of the past.

A distant relative of Mr. Vandeleur, a Limerick banker, through some technical point in the law, took advantage of our president's position, in connection with our society as a manufacturing association and a trading store, and obtained a fiat of bankruptcy against him, which by some legal legerdemain was to protect the interests of his family against the writs in the hands of the sheriff, whose officer was running in hot haste to execute in preference to the claims set up against the estate of a tradesman, storekeeper, and manufacturer. The lawyer's agents made their appearance and urged a demand upon me for all account books, papers, and matters belonging to the estate of the bankrupt. I was in a delemma. There was no tenant right in those days. Our "Agreement" was not accepted as a legal document, nor did I know where to obtain legal advice. There was no Central Board with its legal adviser to offer counsel and guidance. Only one reply came in response to my correspondence, and the advice was "protest against the proceedings." I did protest to the "chiel taking notes," along with two aids or clerks, respectable youths, beside the agent, a man whose dress, face, and head harmonised with his office of evicting tenants. His sallow, mottled face of brass, tinged with a surrounding tint of green, would have rendered it difficult for a painter to obtain suitable pigments to paint his likeness. The only primitive colour on his features was a red-hot hue at the end of his somewhat *retrousse* nose, with apparent latent caloric force enough to drive off rain drops into steam. His small eyes, projecting jaws, and bullet head, served, along with his threadbare clothes —a swallow-tailed coat and breeches—to class him among those who are ever ready to do the dirty work of the bailiffs' court, when evictions are to be executed.

I protested, and he said, "We'll take a note of it." But I saw it was useless, and that his object was only to pacify me to avoid a scene or row before he had completed his inventory. He and his aids went on with their catalogues of all the implements, beginning with the railway truck, which was used to carry the steamed potatoes, &c., from the boiler-house on the railway to the dining-room, and the potato washer, which, like a very large squirrel cage, set in a water box, was a puzzle, and indescribable till he learnt for what use it was made. Then came a list of the horses, cows, corn-ricks, stores, and the school materials to be ready for sale by auction.

LAW IS LAW WITH PRECEDENTS.

As Mr. Vandeleur's family were in favour of the sale, so that the estate and farm might come under the control of Mr. James Molony, of Kiltanon, brother of Mrs. Vandeleur, it was useless contending against the law. I had read Blackstone, and could see the relations between justice and law. But law books and legal practice are often divergent. English law is mainly judge-made law, founded on precedents. Decisions are repeatedly given in accordance with some precedent. The Irish have a wide-spread, strong opinion that English law and lawyers cannot deal justly with the Irish, and this was given to me as a reason why a witness (a woman) had been thrown into a bog hole and drowned. This feeling of doubt of the justice of law, as administered in Ireland, is at the root of much of the agrarian crime that prevails, and the difficulty of obtaining juries to convict.

I have a strong dislike to the terminology of the law books. It is true, law is law—such and so forth, whereby, notwithstanding, aforesaid, nevertheless—at that time it did not recognise the holding of land by an Association of Labourers for a common purpose. The society was without the ægis of the law. Justice may be felt and understood by common men, but not the law. The goddess of Justice is often afflicted with blindness or obliquity of vision; hence the propriety of representing her with a bandage over her eyes.

I had then, and still have, a very strong antipathy or dislike to litigation, and, moreover, there was neither means at hand nor time to avert the evil impending. The members were held to be common labourers, with no rights or claims for improvements, as all they had created and added to the estate belonged to the landlord and his creditors. Legally they were right. It was robbery nevertheless. We had paid our rent, but were remorselessly evicted. We had no remedy. Ruin came upon us suddenly, and Social Co-operation at Ralahine was at an end.

LABOUR NOTES SAVED BY MEMBERS.

Owing to the superiority as well as simplicity of our social arrangements, every member had saved, and accumulated labour notes beyond what each had expended on food, clothing, rent, or lodgings, so that they were not thrown destitute upon the world. The amount of labour notes printed for currency was, nominally, of the value of £50. These notes had been extremely useful as currency for wages, and in their acceptance at the store for food, clothing, and rent of cottages.

The members had amongst them about one-half the amount issued, or £25 ; and, unfortunately, like the so-called "Bank of England," we had issued tokens, or "promises to pay on demand," without having wealth, gold, or deposits to meet these promises or tokens. This was a serious error, both in principle and practice, and fell on me as another heavy penalty in relying on Mr. Vandeleur —to keep faith with those who accepted the notes—to discharge his obligations. There was now, what may be termed, a run on the bank, my resources were soon exhausted, and then panic fear affected those who had not got their notes cashed. The dairy woman and her husband, the herdsman, had saved £5 between them.

I had determined to raise the money and cancel all the notes, to prevent the disappointment and outcry against the "New System" which would have followed if I had left them uncashed, or uncancelled. It was my intention to sell my furniture and the pneumatic apparatus in Limerick to obtain funds, but the agent claimed the air-pump as part of the estate under bankruptcy. No demand was made for the gun, although they had as good a claim for one as for the other. I subsequently got £5 advanced on the gun by Mr. Redman Barry, of Glandore. What object Mr. Vandeleur had in leaving his double-barreled gun with me was a matter of uncertainty. The agent might secretly imagine that in my disappointment and distress I would seek to escape from the hard irony of events, take my own *quietus*, and give them no more trouble.

I obtained assistance, under future responsibilities, and cashed all the outstanding labour notes. The box containing the various kinds of notes showed that every note was redeemed and not a single one was lost or destroyed. These and the whole of the daily labour sheets, showing how every item had been expended, on the farm, the family, and improvements, along with the day book and journal, I handed to the banker at Limerick. The ledger, and other documents, I retained, as necessary in justification of the work done, and for the details given in this history, and without which it could not have been written.

The fact that I had protested against an inventory of the stock and machinery aroused something of the old feeling and prejudice against me among the friends of Mrs. Vandeleur; her brother James Molony, of Kiltanon, was to become the guardian of the children and trustee of Ralahine, and he made my position extremely irksome, unpleasant, and unbearable. I now saw the old suspicions reviving, the people wasting the fruits of their own labour, for it was no longer of any value to them. In the thrashing mill, vast quantities of grain were allowed to go out of the mill along with the straw, to be injured and destroyed by the rain.

My work at Ralahine was done and I prepared for my departure. Desirous of giving expression to their feelings and opinions, as to the benefits they had enjoyed under the new system in their relation with the land and the landlord, the members held a general meeting of the association on the 23rd of November, 1833, when the following declaration was adopted and signed:—

We, the undersigned members of the Ralahine Agricultural and Manufacturing Co-operative Association have experienced for the last two years contentment, peace and happiness under the arrangements introduced by Mr. Vandeleur and Mr. E. T. Craig.

At the commencement we were opposed to the plans

proposed by them; but, on their introduction, we found our condition improved, our wants more regularly attended to, and our feelings towards each other were at once entirely changed from jealousy, hatred, and revenge, to confidence, friendship, and forbearance.

The lectures delivered by the president and Mr. Craig were calculated for our improvement, and have had their influence upon our minds; and the rules formed at the commencement have been very useful in the practical operations of the society.

The subsequent history of Ralahine and county Clare show a marked contrast to the peace and harmony existing under social co-operation, and the conditions that require Peace Preservation Acts to control them.

For many years there was a manifest improvement in the people about Ralahine, but the old antagonisms developed the former bad feelings, and violent passions were engendered. In 1870 Ireland became distracted by agrarian outrages, attended by murders of the foulest character, committed daily. They were rife in County Clare, and perpetrated on property belonging to the Vandeleur family, although in no way referable to the action of the landlord. According to a newspaper account of that day, "A man named Doherty was murdered, near Enistymon, by stabs with a pitchfork, while passing at night by the house of his enemies." It was sufficient for the paper to state that the man held land under Colonel Vandeleur, and had money in the bank. In another paper, at about the same time, there appeared a statement that, "At Crusheen, in the county of Clare, notices have, it is reported, been posted up, commanding the local landlords to reduce their rents 20 per cent., under penalty of being shot. Notice is also given to tenants not to pay the present rents, under a similar penalty. Three landlords are specially named."

In 1881 we find the old evidence repeated. The relations of the labourer to the land and the landlord make peace difficult, if not impossible. County Clare was one of the first to be proclaimed in a state of insurrection, and the Habeas Corpus Act suspended. At the present moment there are more reports of murders in Clare than in any other part of Ireland. On Saturday night, 22nd of October, a tenant farmer, named Patrick Maloney, residing near Ennis, county Clare, was shot while sitting in his parlour after having partaken of supper. Two shots were fired, and Maloney died instantly. Tenants of an estate in Clare are afraid to pay their rents because two agrarian murders have been committed in their neighbourhood.

Another terrible murder is reported from county Clare, two miles from Ennis, in the direction of Enistymon. It appears that Thomas M'Mahon left his own house to visit a neighbour. He did not return, and on the following morning his dead body was found, two miles from his own residence. So frightfully was his head mutilated that identification was difficult. The body had been dragged a considerable distance, and then thrown into the outhouse of a farmer named M'Donnell.

If the reader will contrast county Clare with the terrible condition existing before my labours began, and the present awful state of the country in the neighbourhood of Ralahine, with the happy existence which prevailed among the members of our society, they will be able to appreciate the physical and moral advantages of social organisation over the existing war of opposing interests and passions.

CHAPTER XXXIII.

FAREWELL TO RALAHINE.

My mission in County Clare, although terminated with a suddenness that was painful, had clearly demonstrated the fact that righteousness, even in Ireland, will have its reward, and that a policy of justice and conciliation can effect far better results than one of military force and despotic repression, secured by all the "resources of coercion," which had been almost in constant application since the Union, and yet outrages in the neighbourhood were more numerous than they had ever been known. As already shown, in addition to the assassination of the steward, there were four barbarous murders during the first six weeks after my arrival in the immediate neighbourhood; secret societies were multiplied; and the unemployed turbulent Terry Alts and the White Boys were complete masters of the highways in the district, although coercion was in full force.

From the Union till 1830, nine Coercion Bills had passed, some extending over three years, justifying the statement of Sir Robert Peel, in the debate on Catholic Emancipation in 1829, when he said that "for scarcely a year during the period that has elapsed since the Union, has Ireland been governed by the course of law." From 1830 to 1881, forty-nine coercive measures have been in full swing in Ireland, and the maelstrom of repression is still sustained by a military force of 60,000 men, but which is found insufficient, as crime has greatly increased since the 400 leaders, on mere suspicion, have been thrown into gaol without trial. To aid the force a despicable system of espionage is practised, and subordinate and untrustworthy police agents are employed to tempt the poor and needy by pay, from £10 to £100, for secret information from spies, whose names are to be kept unknown! The Irish character has to a great extent been moulded by English statutes, and this espionage is sowing the poison-fangs of serpents, and the harvest will be an abundant crop of vipers, and cause the death-stings of many innocent persons from retaliatory revenge! How any responsible minister, having a knowledge of the formation of character by education, can sanction such vicious agency is marvellous!

To suppress public meetings in order to silence leaders in the utterance of discontent, and to imprison men by coercion, without trial, is about as prudent and wise as shutting down the steam valve of a steam boiler because the superfluous steam makes a discordant noise. To sit on the safety-valve instead of following Stein's plan of withdrawing the fires—the cause of the heat—is akin to political insanity. The dread as to "separation" is, to me, equally the result of ignorance and fear fed by phantasy. Before my arrival at Ralahine the Terry Alts had had nothing to do, and walked several miles to attend O'Connell's great meetings in favour of Repeal. Some of the labourers I admitted were well known as Terry Alts. Not one of them afterwards attended the large political meetings, and moreover, as I have shown, they were the best workers and the steadiest members of our society. But the end had come.

PAST REMINISCENCES.

In taking farewell of Ralahine every prominent object

seemed to awaken mingled feelings of pleasure, sadness, and regret. The old Castle, with its lofty grey tower, its massive archway, and its wild rocky surroundings, recalled scenes that had been enacted within its blind, windowless walls, from the days of its Milesian chiefs, the tribal septs, and Tanistry, when the bards sang the praises of their princes and the right of the people to the land of their birth, in contrast with the brutal inroads of the grasping and bloody Normans, with their hard, cold, crushing exclusiveness and despotism, ending in the Union, the "English garrison," and its ample supply of Coercion Bills, police espionage, and its crop of poisonous vipers! Ralahine had shown that it was possible to give peace to Ireland without force, by making the people the agents in their own elevation out of poverty and discontent. The contrast effected by the society among the peasantry over the whole of the province was described as immense. The whole of the south of Ireland felt the soothing influence of a system which showed the way to practical independence, and induced in the labourer at Ralahine the content of security and prosperity. After the first two months agrarian outrages began to cease, and during several years were unknown

While Earl Grey's Coercion Bill was in force the number of offences committed in 1833 was 11,444; in 1834 they increased to 14,253! Lord Melbourne had previously declared "captures, special commissioners, and trials useless." During these terrible years of multiplied crimes in defiance of the "resources of coercion," and resistance to rack-rents, and all the horrors of eviction, there were no agrarian outrages in county Clare, nor any murder on the Vandeleur estates till 1870, when the Peace Preservation Act was in force, and a man named Doherty, holding land under the late Colonel Vandeleur, was murdered at Ennistymon with a pitchfork while passing the door of "his enemies." In 1860 there was evidence that the recollection of the blessings of the system at Ralahine still lived in the memory of the people. A traveller thirty miles from Ralahine then heard a number of labourers discussing the condition of the country, and wishing that the "new system" should be tried to heal the social condition of the people.

NOTES.

In writing Notes to Craig's History of Ralahine I have three distinct aims:—(1) To give a brief sketch of the line of thought that led to the forming of the colony; (2) To show that other similar schemes were the outcome of the same wave of thought; and (3) To give an outline of the existing co-operative societies which more or less carry out the ideal of a co-operative agricultural Commonwealth.

The Ralahine Co-operative Experiment was not the result of a sudden impulse of thought, but had been germinating in men's minds for many years. Some of the religious sects which sprang into being at the time of the Reformation had as their economic basis the founding of similar colonies, but their aim was, of course, primarily religious and their secular organization was only the outcome of their religious conceptions. They resembled Ralahine in many ways chiefly in the importance attached to the communal life and in their educational schemes. Thus we find in the Commonwealth of Winstanley the idea of all working to supply the common storehouse and drawing their sustenance from it, and Plockhoy (1659) taught the establishment of "Little Commonwealths." Plockhoy's scheme was developed by the Quaker, Bellers, who proposed the starting of agricultural colonies of about 300 persons, and in industrial districts trading colonies of the same size, which were to work co-operatively and to lead a communistic life. The standard of value was to be labour, not money, which is a feature of Owen's experiments and Ralahine.

The ideal of a communal life was not confined to the Protestant sects, for by far the most successful attempt was made by the Jesuits, who, in 1588 founded their community in Paraguay. The government of that community, however, was directly under the control of the most disciplined body of men the world has ever seen. The community covered a large area, and included many towns, each of which was governed by two Jesuits, assisted by a body of Guarani (as the natives were called) officials. Discipline was carried out in the most extreme manner; each town was as regular and similar as a Roman camp, and the dress of the inhabitants was uniform and prescribed. The Jesuits realised that they must provide for the recreation as well as for the support of their people, and on religious and other festivals organised processions and other pageants.

Property, or rather the produce of the soil, since each house was similar and dress uniform, was divided into three parts. One went to the general use of the community, one to the heads of families and one to the Church. The community built houses, and ships, and carried on trade. The Spaniards of the surrounding districts disliked the community, largely because it was more for the benefit of the natives than of settlers, but the community was so successful that even its enemies were forced to trade with it. It was finally broken up in 1768, largely owing to the hostility of the other European settlers who wished to exploit South America for their own benefit.

The idea is an old one, but it was in the mind of the French commercial traveller, Fourier, that it first took such concrete form as directly to influence a large body of thought. Fourier, who was perhaps the most unpractical of visionaries, spent the whole of his life advocating the establishment of co-operative colonies or "Phalansteries," as he called them. In Fourier's colonies each man was to choose the form of labour that most appealed to him, but labour must be made so attractive that all would voluntarily devote themselves to it. The chief feature of his system was the communal dwelling, which was to be a large building with sets of private apartments. His dislike of fresh air was shown by his grouping his factories, for each colony was to have two or more factories, around his main building and providing covered passages to work, and covered courts for recreation.

Fourier was a contemporary of Robert Owen, the founder of English co-operation, and his ideas had a profound influence on Owen. The following extract from Owen's Report to the Parliamentary Committee on the Poor Laws in 1817 is of interest in this connection. Having laid down the proposition that the best way of dealing with the pauper question was to set up colonies of from 500 to 1,500 persons, he submitted a plan, or rather a sketch, of a country-side with large groups of buildings scattered over it at considerable distances apart. "They are intended," he says, "to represent square buildings, sufficient to accommodate 1,200 persons . . . . within the square are public buildings which divide it into two paallelograms. The central building contains a public kitchen, mess rooms, and all the accommodation necessary to economical and comfortable cooking and eating. To the right of this is a building of which the ground floor will form the infant school and the other a lecture room and place of worship.

" The building to the left contains a school for the older chlldren and a committee room on the ground floor ; above a library and a room for adults. In the vacant space within the square are enclosed grounds for exercise and recreation ; these enclosures to have trees planted in them."

It is not necessary to continue the quotation which goes into an elaborate description of the buildings. The main point that emerges is that in Owen's scheme of a " colony," the ideas of Fourier are clearly discernible. He follows Fourier even further, in that he interferes with family life by taking the children from their parents at three years old and making them sleep in dormitories, and as soon as they are old enough to work they are trained to assist in gardening and manufacture. Like the Phalansteries, Owen's colonies were to have factories attached to them, in which those members of the community not employed in agriculture were to work.

Fourier's idea fertilised two distinct fields : the first, represented by Owen, which came near to success in Ralahine, the second by Godin, who founded a similar society on a purely manufacturing basis. We should first see how Owen's plans developed, and then look at the other branch. In 1825 Owen purchased a tract of land in Indiana, U.S.A., of 30,000 acres, which had been occupied by a sect of German religious fanatics, and established his colony of " New Harmony." Here he attempted to found a colony on a communal basis and in three months 900 persons joined him. No effort was made to select the colonists, but all comers were taken. The first rulers of the Commonwealth made some attempt at maintaining order, but from the account in Mr. Holyoake's " History of Co-operation," the only thing we gather is that they prohibited the sale of intoxicating liquors. This regulation seems to have obscured Mr. Holyoake's view of the community since the greater part of his account is taken up by a panegyrc of temperance and a lurid picture of the evils that follow from drink. The New Harmony colony dragged on an inglorious existence for a couple of years, though hotly attacked on religious grounds by the clergy of all sects, but the population gradually dispersed and those who remained returned to individualism. Owen was connected with other experiments in co-operative communities—Orbiston in Scotland and Queenswood in England.

The Orbiston community preceded Ralahine, as it was founded in 1825, but it only lasted until 1827. The Queenswood experiment came after Ralahine and lasted from 1839 to 1844. In both cases bad management and extravagance caused their failure. The influence of Fourier's ideal Phalanstery with its large halls and buildings seems to have taken hold of the early experimenters and encouraged them to undertake vast building operations with disastrous results.

This short summary gives us a background on which to view the Ralahine experiment.

The other field which Fourier fertilised was the field of industry, and here his chief disciple was Godin, the ironmaster of Guise. Godin, like Owen, began life as a simple worker, but by his industry rose to be proprietor of one of the largest iron works of France. He was born in 1817, just at the time when Fourier's teaching was at its height and Owen was writing his views on the new order of society. Like Owen, he burnt his fingers in attempting to found a commune in the United States. About 1856 he conceived the idea of turning his own works into a communistic society and by 1861 had opened the first part of the "Familistère", of Guise. Godin's main conception in its purely mechanical part, like Owen, followed Fourier. He was a yet more slavish follower; we find the large building with its private apartments, the roofed-in courts, the covered passages. Godin tried to reproduce a Phalanstery as closely as was possible in a community depending on a single industry. His original conception was that all his workers should live in the communal building and that they should live a somewhat communal life, but he found that numbers of his men refused to join the full community and he had to modify his scheme accordingly.

Godin, though an idealist, was exceedingly practical in his methods. He realised that were he to hand over his business *en bloc* to the workers they would be most unlikely to be able to manage it and, therefore, he invented an ingenious scheme whereby the workers would gradually acquire control and at the same time learn how to manage the business. He divided his men into four classes. 1. *Auxiliaries*, which included new hands, casual workers, etc. 2. *Participants*, who receive one share each of the profits calculated on their wages. 3. *Sociétaires*, who receive a share and a half of the profits. 4. *Associés*, who receive a double share. In 1907 there were 422 of class 4, 71 class 3, 801 class 2, and 907 class 1. The Associés are a picked band and fill up all vacancies that arise in their ranks from the Sociétaires. They elect the committee of management of the works and generally supervise the running of the business and the community. They must live in the Familistère. The system worked admirably and has attached to it elaborate systems of insurance of members, education of children, etc. There is a nursery in which members can leave their children should they be unable to look after them themselves. There is also a theatre, recreation hall, gymnasium, etc.

The actual handing over of the business to the workers was worked by the profit-sharing system. The workers' share of the profits was distributed in shares, which gradually bought out Godin's interest in the business, a process which was accelerated by his leaving a large share to the society on his death. By 1895 the society owned the whole business.

Godin's work survived the death of its founder, and was flourishing at the outbreak of the late war. A branch of the business was opened near Brussels. During the war both the factories were occupied by the Germans and suffered greatly; the machinery was destroyed but a devoted band of the old workers have re-established themselves there and have re-opened on a small scale. It is much to be hoped that the community will soon regain its old prosperity.

Godin was not alone in following Fourier's ideas in the industrial world. LeClaire in Paris set up a self-governing body of workers in the builders' decorating business and there are several co-operative productive societies in France, chiefly of builders, though there are also other works such as glass workers' shops and even cab drivers' societies.

Thus, the idea which gave birth to Ralahine developed itself. In agriculture it seemed to have failed, while in industry it had achieved some notable successes.

At the end of the nineteenth century a new impulse was given the movement in Italy where a very different set of circumstances led to the establishment of collective farming societies which, though they do not approach Ralahine in completeness, shew that the co-operative system of farming can be successful.

The movement began about 1890 and was the result of the trade-union activities of the Italian agricultural labourers. The trade unions became very powerful and organised a number of serious strikes which resulted in a great increase of wages. The farmers, however, met the increase in wages by altering their system of farming, giving up rice growing in favour of fattening cattle and introducing machinery to replace their men. As a result of this there was a great deal of unemployment among the agricultural labourers, though the conditions of those in work were better than before. In the early years of the twentieth century the crisis grew acute and the labourers' unions rented land with a view to giving employment to their out-of-work members. It was in the province of Reggio Emilia that the movement developed most and by 1912 there were 69 collective farms in Italy.

The usual system is for a society to rent land on a lease of from 9 to 12 years. The Italian system of land tenure lends itself to such a system as a large part of the land of Italy is held on the métayer system, which provides that the landlord shall supply stock, seeds, etc., and the tenant in return pay a proportion of the proceeds or crop. It is, however, necessary for the tenant to provide security for his rent, and this in the case of the collective farming societies was often done by the co-operative stores. There are other systems of land holding in Italy, and the method adopted by the society depends largely on the way in which its land is held. In some cases it is purchased outright by the society.

The first money of the society is raised by share capital. Each member is compelled to buy a certain number of shares. The shares may be from 5 lire to 40 lire, but usually only a small portion is paid on entrance and the remainder in weekly or monthly instalments. Members' shares in the distribution of profits are usually devoted to liquidating the unpaid portions of their shares.

There are two different plans on which the societies work: (1) The land is divided among the members and farmed by them as individuals; (2) The land is farmed in common by the members under the direction of a manager appointed by the society. The first system is not in any way comparable to the Ralahine experiment but it has met an economic difficulty which may well arise in Ireland where the position of the agricultural labourer is becoming more and more like that of the Italian labourer of twenty years ago. The trade union movement is spreading very rapidly among the Irish labourers and there are indications that attempts will be made to secure land for them. Therefore, the question of establishing colonies of labourers, each owning and managing part of a large tract of land, is of importance in Ireland and Western Europe generally. But the admirers of Ralahine will be more interested in the second form.

The second form is more purely co operative since the members actually work the land in common. The societies usually have more members than are necessary to cultivate the soil but the extra members work elsewhere as labourers for hire. All the members of the society belong to the labourers' union and the union employment bureau assigns them to their jobs.

The society is governed by the general meeting of the members usually held once a year. It elects a committee which in turn chooses a president and secretary. The committee which meets at frequent intervals performs all the functions of business management, contracts, loans, etc. When necessary it chooses a business manager. There are also supervising and arbitration provisions. The societies are generally constituted with limited liability.

The provisions as to the disposal of profits vary, but usually a large proportion is devoted to the collective funds of the Society and in some cases actual deductions are made from the members' wages during the early years of the Society's life in order to increase those funds. A proportion goes to reserves, and the remainder is divided among the members in proportion to the work done.

The following extracts from the rules of the Co operative Farming Society at Altedo will serve as an example of how this form of Society is constituted.

The objects are wide and are given in full.

" RULE 4.

(a) To undertake the carrying out of agricultural work and management of rural properties whose cultivation will be entrusted to the members.

(b) To carry on trade in agricultural produce, and in all articles necessary to the industry of agriculture.

(c) To purchase agricultural instruments and machinery.

(d) To purchase lands, purchase and erect buildings necessary for working farm properties and for their own members to live in.

(e) To open one or more shops for the purpose of supplying the members with their requirements at cost price.

(f) To spread agricultural education among the members by means of lectures and the distribution of pamphlets.

(g) To combine in the insurance of members in the national societies against accidents and sickness."

It will thus be seen that the objects are very wide and include all those of Ralahine except the education of children. Indeed, the first half of the third rule which comes before the specified objects just given might have been written by one of the early Nineteenth Century reformers who founded associations "For the regeneration of mankind and the changing of the whole moral system," etc. It reads: "The Society undertakes the continuous policy of ameliorating the economic and moral conditions of farm labourers . . . ."

Share capital is unlimited in amount and the shares are value 10 lire each (8s. 4d.). The shares are untransferable, but, if a member resigns, all payments made by him on account of shares are repaid to him less his share of any losses suffered by the Society within the year, and in case of death his heirs may nominate one of their number to represent him.

Membership is open to all farm servants and day labourers resident in Altedo of not less than 16 years old. Each member is bound to pay an entrance fee of one lire and to buy at least one share which may be paid for at once or in monthly instalments of 50 centessimi.

Each member has the right to be admitted to a share in all work undertaken by the Society and to share in the profits in proportion to the number of days' work done by him during the year.

The balance sheet is presented annually and the profits divided as follows:—

(*a*) 20 per cent. to the Reserve Fund.

(*b*) 40 per cent. to the Collective Fund.

(*c*) 40 per cent. to the members in proportion to work done, provided that no profits are divided until the Collective Fund has reached one-tenth of the total of the share capital and Reserve Fund. The Collective Fund is distinct from the Reserve and is to be used for furthering the aims of the Society, especially (*c*), (*d*), (*e*), and (*f*).

The above is a summary of the chief rules of the Altedo Society.

In 1916 seven collective farms in the province of Reggio Emilia (this excludes all farms in which the tenancy is collective, but are worked by the members as individuals), had a total membership of 2,012 and farmed 1,052 hectares, or about 2,650 acres. They paid rent amounting to £4,640. The oldest Society was formed in 1890, and the youngest in 1911. In the province of Ravenna in 1917 there were 30 societies, but there was some over-lapping, as the Republican and Socialist parties each developed its own farms. The total number of members was 9,524, and they farmed about 13,000 acres. In Bologna 12 Societies had 2,790 members and farmed about 4,580 acres.

There are also societies in many other parts of Italy, and in Sicily, where the system of landholding is somewhat different. In Sicily, in 1915, there were 40 societies, with about 70,000 acres of land, but all these societies merely hold the land in common; it is worked by the members as individuals.

The results obtained have been satisfactory. According to the International Review of Agricultural Economics for August, 1918, "the conversion of labourers into farmers effected by the agency of association, has procured for them above all, a more exact knowledge of the means of agriculture as regards both the cost and the productivity of these means. It has given birth among them to a more accurate sense of their own responsibility. It has ensured to them, and especially to the labourers of Sicily, who are still bound by some feudal customs, a greater freedom of thought and activity.

"The associated labourers have not always succeeded in obtaining for their work higher pay than they received before the collective farming was initiated. But it is certain that in many districts this method of holding land has taken an effective part in reducing or suppressing unemployment and emigration, in that it has distributed the profits of farming among a larger number of persons. In many cases it has contributed to the realization of true progress."

As might be expected, the chief difficulties are in the management and book-keeping departments, but the "Bureau of Inspection and Assistance for Co-operative Farming" which has been formed by the "National Institute of Co-operative Credit" has the object of helping to remedy these defects.

Altogether the example of the Italian co-operative farms is most encouraging. Roumania also has co-operative societies for leasing land which is cultivated in separate allotments by the members, and in Serbia there were societies of both types. The English "Allotment Societies" resemble the Roumanian and Italian collective leasing societies, but there is not any example that I know of of collective cultivation in England.

In Ireland there is one example at Foynes, Co. Limerick. The area farmed was only seven acres and the results have only just justified the experiment. An article by Mr. E. E. Lysaght in the November number of the Irish Monthly is, however, very suggestive, and, if he is able to carry out his idea, it will mean that Co. Clare is again leading the world in the collective working of land. Mr. Lysaght holds that, in order to make a scheme of collective ownership effectual it is desirable that the workers should first of all combine in running part of the farm or industry, and as they show capacity and accumulate funds by their united effort, they should take over more and more branches until they control the whole. Should his ideas be carried into practice it will be a very instructive experiment for the peoples of North-Western Europe.

D: OC.

The Co-Operative Reference Library.
Dublin, December, 1919.

Appendix I

AN IRISH UTOPIA

JAMES CONNOLLY

"Were the hand of Locke to hold from heaven a scheme of government most perfectly adapted to the nature and capabilities of the Irish nation, it would drop to the ground a mere sounding scroll were there not other means of giving it effect than its intrinsic excellence. All true Irishmen agree in what ought to be done, but how to get it done is the question."—*Secret Manifesto* (Ireland), 1793.

IN our last chapter we pointed out how the close of the Napoleonic wars precipitated a commercial crisis in Great Britain and Ireland, and how in the latter country it also served to intensify the bitterness of the relations existing between landlord and tenant. During the continuation of the wars against Napoleon, agricultural prices had steadily risen owing to the demand by the British Government for provisions to supply its huge army and navy. With the rise in prices rents had also risen, but when the close of the war cut off the demand, and prices consequently fell, rents did not fall along with them. A falling market and a stationary or rising rent-roll could have but one result in Ireland—viz., agrarian war.

The landlords insisted upon their "pound of flesh", and the peasantry organised in secret to terrorise their oppressors and protect themselves. In the year 1829 a fresh cause of popular misery came as a result of the Act granting Catholic Emancipation. Until that year no Catholic had the right to sit in the English House of Commons, to sit on the Bench as a Judge, or to aspire to any of the higher posts in the Civil, Military, or Naval services. As the culmination of a long fight against this iniquitous "Protestant Ascendancy", after he had aroused the entire Catholic population to a pitch of frenzy against the injustices inherent in it, the Catholic leader, Daniel O'Connell, presented himself as a candidate for the representation in Parliament of the County Clare, declaring that if elected he would refuse to take the oath then required of a Member of Parliament, as it libelled the Catholic Religion. In Ireland at that time open voting prevailed, every elector having to openly declare before the clerks of the election and all others who chose to attend, the name of the candidate for whom he voted. In Ireland at that time also most of the tenants were tenants-at-will, removable at the mere pleasure of the agent or landlord. Hence elections were a combination of farce and tragedy—a farce as far as a means of ascertaining the real wish of the electors were concerned, a tragedy whenever any of the tenants dared to vote against the nominee of the landlord. The suffrage had been extended to all tenants paying an annual rental of forty shillings, irrespective of religious belief, but the terrible power of life and death possessed by the landlord made this suffrage ordinarily useless for popular purposes. Yet when O'Connell appealed to the Catholic peasantry of Clare to brave the vengeance of their landed tyrants and vote for him in the interests of religious liberty they nobly responded. O'Connell was elected, and as a result Catholic Emancipation was soon afterwards achieved. But the ruling classes and the British Government took their revenge by coupling with this reform a Bill depriving the smaller tenants of the suffrage, and raising the amount of rent necessary to qualify for a vote to ten pounds.

Up till that time landlords had rather encouraged the growth of population on their estates, as it increased the number of their political adherents, but with the passage of this Act of Parliament this reason ceased to exist and they immediately began the wholesale eviction of their tenantry and the conversion of the arable lands into grazing farms. The Catholic middle, professional and landed class by Catholic Emancipation had the way opened to them for all the snug berths in the disposal of the Government; the Catholics of the poorer class as a result of the same Act were doomed to extermination to satisfy the vengeance of a foreign Government, and an aristocracy whose power had been defied where it knew itself most supreme.

The wholesale eviction of the smaller tenants and the absorption of their farms into huge grazing ranches, thus closing up every avenue of employment to labour, meant death to the agricultural population, and hence the peasantry struck back by every means in their power. They formed lodges of the Secret Ribbon Society, made midnight raids for arms upon the houses of the gentry, assembled at night in large bodies and ploughed up the grass lands, making them useless for grazing purposes, filled up ditches, terrorised graziers into surrendering their ranches, wounded and killed those who had entered the service of graziers or obnoxious landlords, assassinated agents, and sometimes in sheer despair opposed their unarmed bodies to the arms of military. Civil war of the most sanguinary character was convulsing the country; in May, 1831, the Lord Lieutenant of Ireland and a huge military force accompanied by artillery marched through Clare to overawe the people, but as he did not stop evictions, nor provide employment for the labourers whom the establishment of grazing had deprived of their usual employment on the farm, the "outrages" still continued. Nor were the professional patriots, or the newly emancipated Catholic rich, any more sympathetic to the unfortunate people. They had opened the way for themselves to place and preferment by using the labourer and cottier-farmer as a lever to overthrow the fortress of religious bigotry and ascendancy, and now when the fight was won they abandoned these poor co-religionists of theirs to the tender mercies of their economic masters. To the cry of despair welling up from the hearts of the evicted families, crouching in hunger upon the road-side in sight of their ruined homes, to the heartbroken appeal of the labourer permanently disemployed by the destruction of his source of employment; to the wail of famishing women and children the politicians invariably had but one answer—"Be law abiding, and wait for the Repeal of the Union". We are not exaggerating. One of the most ardent Repealers and closest friends of Daniel O'Connell, Mr. Thomas Steele, had the following manifesto posted up in the Market Place of Ennis and other parts of Clare, addressed to the desperate labourers and farmers:

"Unless you desist, I denounce you as traitors to the cause of the liberty of Ireland . . . I leave you to the Government and the fire and bayonets of the military. Your blood be upon your own souls."

This language of denunciation was uttered to the heroic men and women who had sacrificed their homes, their security, and the hopes of food for their children to win the emancipation from religious tyranny of the well-fed snobs who thus abandoned them. It is difficult to see how a promised Repeal of the Union some time in the future could have been of any use to the starving men of Clare, especially when they knew that their fathers had been starved, evicted and tyrannised over *before* just as they were *after* the Union. At that time, however, it was deemed a highly patriotic act to ascribe all the ills that Irish flesh is heir to the Union. For example, a Mr. O'Gorman Mahon, speaking in the House of Commons, London, on February 8, 1831, hinted that the snow storm then covering Ireland was a result of the Legislative Union. He said:

"Did the Hon. Members imagine that they could prevent the unfortunate men who were under five feet of snow from thinking they could better their condition by a Repeal of the Union. It might be said that England had not caused the snow, but the people had snow on them, and they thought that their connection with England had reduced them to the state in which they now were."

Another patriot, destined in after years to don the mantle of an Irish rebel, William Smith O'Brien, at this time, 1830, published a pamphlet advocating emigration as the one remedy for Irish misery.

On the other hand, a Commission appointed by the House of Lords in 1839 to inquire into the causes of the unrest and secret conspiracies amongst the poorer class examined many witnesses in close touch with the life of the peasantry and elicited much interesting testimony tending to prove that the evil was much more deeply rooted than any political scheme of Government, and that its real roots were in the social conditions. Thus examined as to the attitude of the labourers towards the Ribbon Association, one witness declared:

"Many look to the Association for protection. They think they have no other protection."

Question: "What are the principal objects they have in view?"

Answer: "To keep themselves upon their lands. I have often heard their conversation, when they say:

" 'What good did Emancipation do for us? Are we better clothed or fed, or are our children better clothed and fed? Are we not as naked as we were, and eating dry potatoes when we can get them? Let us notice the farmers to give us better food and better wages, and not give so much to the landlord, and more to the workman; we must not be letting them be turning the poor people off the ground.' "

And a Mr. Poulett Scroope, M.P., declared in one of his writings upon the necessity for a Poor Law: "The Tithe question, the Church, the Grand Jury laws, the more or fewer Catholics appointed to the Sheriffalty or Magistracy—these are all topics for political agitation among the idle mobs; but the midnight massacre, the daily plunder, the frequent insurrection, the insecurity of life and property throughout agricultural districts of Ireland, these are neither caused by agitation, nor can be put down with agitation."

It will be thus seen that the opinion of the independent Member of Parliament coincided with that of the revolting labourers as to the relative unimportance to the toilers of Ireland of the subjects which then as now bulked most largely in the minds of politicians.

This was the state of things political and social in Ireland in the year 1831, and as it was in Clare the final effective blow had been struck for religious emancipation so it also was Clare that was destined to see the first effort to discover a peaceful way of achieving that social Emancipation, without which all other freedom, religious or political, must ever remain as Dead Sea fruit to the palate of Labour.

In 1823 the great English socialist, Robert Owen, visited Ireland and held a number of meetings in the Rotunda, Dublin, for the purpose of explaining the principles of Socialism to the people of that city. His audiences were mainly composed of well-to-do inhabitants, as was, indeed, the case universally at that period when Socialism was the fad of the rich instead of the faith of the poor. The Duke of Leinster, the Catholic Archbishop Murray, Lord Meath, Lord Cloncurry and others occupied the platform, and as a result of the picture drawn by Owen of the misery then existing, and the attendant insecurity of life and property amongst all classes, and his outline of the possibilities which a system of Socialist co-operation could produce, an association styling itself the "Hibernian Philanthropic Society" was formed to carry out his ideas. A sum of money was subscribed to aid the prospects of the society, a General Brown giving £1,000, Lord Cloncurry £500, Mr. Owen himself subscribing £1,000, and

£100 being raised from other sources. The society was short lived and ineffectual, but one of the members, Mr. Arthur Vandeleur, an Irish landlord, was so deeply impressed with all he had seen and heard of the possibilities of Owenite Socialism, that in 1831, when crime and outrage in the country had reached its zenith, and the insecurity of life in his own class had been brought home to him by the assassination of the steward of his estate for unfeeling conduct towards the labourers, he resolved to make an effort to establish a Socialist colony upon his property at Ralahine, County Clare. For that purpose he invited to Ireland a Mr. Craig of Manchester, a follower of Owen, and entrusted him with the task of carrying the project into execution.

Though Mr. Craig knew no Irish, and the people of Ralahine, as a rule, knew no English—a state of matters which greatly complicated the work of explanation—an understanding was finally arrived at, and the estate was turned over to an association of the people organised under the title of "The Ralahine Agricultural and Manufacturing Co-Operative Association".

In the preamble of the Laws of the Association, its objects were defined as follows:

"The acquisition of a common capital.

"The attainment of a greater share of the comforts of life of poverty, sickness, infirmity, and old age.

"The attainment of a greater share of the comforts of life than the working classes now possess.

"The mental and moral improvement of its adult members.

"The education of their children."

The following paragraphs selected from the Rules of the Association will give a pretty clear idea of its most important features:

"BASIS OF THE SOCIETY

"That all the stock, implements of husbandry, and other property belong to and are the property of Mr. Vandeleur,

AN IRISH UTOPIA

until the society accumulates sufficient to pay for them; they then become the joint property of the Society.

" Production

"We engage that whatever talents we may individually possess, whether mental or muscular, agricultural, manufacturing, or scientific, shall be directed to the benefit of all, as well by their immediate exercise in all necessary occupations as by communicating our knowledge to each other, and particularly to the young.

"That, as far as can be reduced to practice, each individual shall assist in agricultural operations, particularly in harvest, it being fully understood that no individual is to act as steward, but all are to work.

"That all the youth, male or female, do engage to learn some useful trade, together with agriculture and gardening, between the ages of nine and seventeen years.

"That the committee meet every evening to arrange the business for the following day.

"That the hours of labour be from six in the morning till six in the evening, in summer, and from daybreak till dusk, in winter, with the intermission of one hour for dinner.

"That each agricultural labouring man shall receive eightpence, and every woman fivepence per day for their labour (these were the ordinary wages of the country, the secretary, storekeeper, smiths, joiners, and a few others received something more; the excess being borne by the proprietors) which it is expected will be paid out at the store in provisions or any other article the society may produce or keep there; any other articles may be purchased elsewhere.

"That no member be expected to perform any service or work but such as is agreeable to his or her feelings, or they are able to perform; but if any member thinks that any other member is not usefully employing his or her time, it is his or her duty to report it to the committee, whose duty it will be to bring that member's conduct before a general meeting, who shall have power, if necessary, to expel that useless member.

" Distribution and Domestic Economy

"That all the services usually performed by servants be performed by the youth of both sexes under the age of seventeen years, either by rotation or choice.

"That the expenses of the children's food, clothing, washing, lodging, and education be paid out of the common funds of the society, from the time they are weaned till they arrive at the age of seventeen, when they shall be eligible to become members.

"That a charge be made for the food and clothing, etc., of those children trained by their parents, and residing in their dwellinghouses.

"That each person occupying a house, or cooking and consuming their victuals therein, must pay for the fuel used.

"That no charge be made for fuel used in the public room.

"That it shall be a special object for the sub-committee of domestic economy, or the superintendent of that department to ascertain and put in practice the best and most economical methods of preparing and cooking the food.

"That all the washing be done together in the public wash-house; the expenses of soap, labour, fuel, etc., to be equally borne by all the adult members.

"That each member pay the sum of one half-penny out of every shilling received as wages to form a fund to be placed in the hands of the committee, who shall pay the wages out of this fund to any member who may fall sick or meet with an accident.

"Any damage done by a member to the stock, implements, or any other property belonging to the society to be made good out of the wages of the individual, unless the damage is satisfactorily accounted for to the committee.

" EDUCATION AND FORMATION OF CHARACTER

"We guarantee each other that the young children of any person dying, whilst a member of this society, shall be equally protected, educated and cherished with the children of the living members, and entitled, when they arrive at the age of seventeen, to all the privileges of members.

"That each individual shall enjoy perfecty liberty of conscience, and freedom of expression of opinion, and in religious worship.

'That no spirituous liquors of any kind, tobacco, or snuff be kept in the store, or on the premises.

"That if any of us should unfortunately have a dispute with any other person, we agree to abide by a decision of the majority of the members, or any person to whom the matter in question may be by them referred.

"That any person wishing to marry another do sign a declaration to that effect one week previous to the marriage taking place, and that immediate preparations be made for the erection, or fitting up of a suitable dwellinghouse for their reception.

"That any person wishing to marry another person, not a member, shall sign a declaration according to the last rule; the person not a member shall then be balloted for, and if rejected, both must leave the society.

"That if the conduct of any member be found injurious to the well-being of the society, the committee shall explain to him or her in what respect his or her conduct has been injurious, and if the said member shall continue to transgress the rules, such member shall be brought before a general meeting, called for the purpose, and if the complaint be substantiated, three fourths of the members present shall have power to expel, by ballot, such refractory member.

" GOVERNMENT

"The Society to be governed, and its business transacted, by a committee of nine members, to be chosen half-yearly, by ballot, by all the adult male and female members, the ballot list to contain at least four of the last committee.

"The committee to meet every evening, and their transactions to be regularly entered into a minute book, the recapitulation of which is to be given at the society's general meeting by the secretary.

"That there be a general weekly meeting of the society; that the treasurer's accounts be audited by the committee, and read over to the society; that the 'Suggestion Book' be also read at this meeting."

The colony did not use the ordinary currency of the country, but instead adopted a "Labour Note" system of payment, all workers being paid in notes according to the number of hours worked, and being able to exchange the notes in the store for all the necessities of life. The notes were printed on stiff cardboard about the size of a visiting card, and represented the equivalent of a whole, a half, a quarter, an eighth, and a sixteenth of a day's labour. There were also special notes printed in red ink representing respectively the labours of a day and a half, and two days. In his account of the colony published under the title of "History of Ralahine", by Heywood & Sons, Manchester (a book we earnestly recommend to all our own readers), Mr. Craig says: "The labour was recorded daily on a 'Labour Sheet', which was exposed to view during the following week. The members could work or not at their own discretion. If no work, no record, and, therefore, no pay. Practically the arrangement was of great use. There were no idlers." Further on he comments:

"The advantages of the labour notes were soon evident in the saving of members. They had no anxiety as to employment, wages, or the price of provisions. Each could partake of as much vegetable food as he or she could desire. The expenses of the children from infancy, for food or education, were provided for out of the common fund.

"The object should be to obtain a rule of justice, if we seek the law of righteousness. This can only be fully realised in that equality arising out of a community of property where the labour of one member is valued at the rate as that of another member, and labour is exchanged for labour. It was not possible to attain to this condition of equality at Ralahine, but we made such arrangements as would impart a feeling of security, fairness and justice to all. The prices of provisions were fixed and uniform. A labourer was charged one shilling a week for as many vegetables and as much fruit as he chose to consume; milk was a penny per quart; beef and mutton fourpence, and pork two and one-half pence per pound. The married members occupying separate quarters were charged sixpence per week for rent, and twopence for fuel."

In dealing with Ireland no one can afford to ignore the question of the attitude of the clergy; it is therefore interesting to quote the words of an English visitor to Ralahine, a Mr. Finch, who afterwards wrote a series of fourteen letters describing the community, and offered to lay a special report before a Select Committee of the House of Commons upon the subject. He says:

"The only religion taught by the society was the unceasing practice of promoting the happiness of every man, woman, and child to the utmost extent in their power. Hence the Bible was not used as a school-book; no sectarian opinions were taught in the schools; no public dispute about religious dogmas or party political questions took place; nor were members allowed to ridicule each other's religion; nor were there any attempts at proselytism. Perfect freedom in the performance of religious duties and religious exercises was guaranteed to all. The teaching of religion was left to ministers of religion and to the parents; but no priest or minister received anything from the funds of the society. Nevertheless, both Protestant and Catholic priests were friendly to the system as soon as they understood it, and one reason was they found these sober, industrious persons had now a little to give them out of their earnings, where as formerly they had been beggars."

Mr. Craig also states that the members of the community after it had been in operation for some time, were better Catholics than before they began. He had at first considerable difficulty in warding off the attacks of zealous Protestant proselytisers, and his firmness in doing so was one of the chief factors in winning the confidence of the people as well as their support in insisting upon the absolutely non-sectarian character of the teaching.

All disputes between the members were settled by appeals to a general meeting in which all adults of both sexes participated, and from which all judges, lawyers, and other members of the legal fraternity were rigorously excluded.

To those who fear that the institution of common property will be inimical to progress and invention, it must be reassuring to learn that this community of "ignorant" Irish peasants introduced into Ralahine the first reaping machine used in Ireland, and hailed it as a blessing at a time when the gentlemen farmers of England were still gravely debating the practicability of the invention. From an address to the agricultural labourers of the County Clare, issued by the community on the occasion of the introduction of this machine, we take the following passages, illustrative of the difference of effect between invention under common ownership and under capitalist ownership:

"This machine of ours is one of the first machines ever given to the working classes to lighten their labour, and at the same time to increase their comforts. It does not benefit any one person among us exclusively, nor throw any individual out of employment. Any kind of machinery used for shortening labour—except used in a co-operative society like ours—must tend to lessen wages, and to deprive working men of employment, and finally either to starve them, force them into some other employment (and then reduce wages in that also) or compel them to emigrate. Now, if the working classes would cordially and peacefully unite to adopt our system, no power or party could prevent their success."

This was published by order of the committee, 21st August, 1833, and when we observe the date we cannot but wonder at the number of things Clare—and the rest of Ireland—has forgotten since.

It must not be supposed that the landlord of the estate on which Ralahine was situated had allowed his enthusiasm for Socialism to run away with his self-interest. On the contrary, when turning over his farms to the community he stipulated for the payment to himself of a very heavy rental in kind. We extract from "Brotherhood", a Christian Socialist Journal published in the North of Ireland in 1891, a statement of the rental, and a very luminous summing up of the lesson of Relahine, by the editor, Mr. Bruce Wallace, long a hard and unselfish worker for the cause of Socialism in Ireland:

"The Association was bound to deliver annually, either at Ralahine, Bunratty, Clare, or Limerick, as the landlord might require, free of expense—

Wheat	320 brls.
Barley	240 brls.
Oats	50 brls.
Butter	10 cwt.
Pork	30 cwt.
Beef	70 cwt.

"At the prices then prevailing, this amount of produce would be equivalent to about £900, £700 of rent for the use of natural forces and opportunities, and £200 of interest upon capital. It was thus a pretty stiff tribute that these poor Irish toilers had to pay for the privilege of making a little bit of their native soil fruitful. This tribute was, of course, so much to be deducted from the means of improving their sunken condition. In any future efforts that may be made to profit by the example of Ralahine and to apply again the principles of co-operation in farming, there ought to be the utmost care taken to reduce to a minimum the tribute payable to non-workers, and if possible to get rid of it altogether. If, despite this heavy burden of having to produce a luxurious maintenance for loungers, the condition by the toilers at Ralahine, as we shall see, was marvellously raised by the introduction of the co-operative principle amongst them, how much more satisfactorily would it have been raised had they been free of that depressing dead weight?"

Such is the lesson of Ralahine. Had all the land and buildings belonged to the people, had all other estates in Ireland been conducted on the same principles, and the industries of the country also organised, had each of them appointed delegates to confer on the business of the country at some common centre as Dublin, the framework and basis of a free Ireland would have been realised. And when Ireland does emerge into complete control of her own destinies she must seek the happiness of her people in the extension on a national basis of the social arrangements of Ralahine, or else be but another social purgatory for her poor—a purgatory where the pangs of the sufferers will be heightened by remembering the delusive promises of political reformers.

In the most crime-ridden county in Ireland this partial experiment in Socialism abolished crime; where the fiercest fight for religious domination had been fought it brought the mildest tolerance; where drunkenness had fed fuel to the darkest passions it established sobriety and gentleness; where poverty and destitution had engendered brutality, midnight maruading, and a contempt for all social bonds, it enthroned security, peace and reverence for justice, and it did this solely by virtue of the influence of the new social conception attendant upon the institution of common property bringing a common interest to all. Where such changes came in the bud, what might we not expect from the flower? If a partial experiment in Socialism, with all its drawbacks of an experiment, will achieve such magnificent results what could we not rightfully look for were all Ireland, all the world, so organised on the basis of common property, and exploitation and mastership forever abolished?

The downfall of the Association came as a result of the iniquitous land laws of Great Britain refusing to recognise the right of such a community to hold a lease or to act as tenants. The landlord, Mr. Vandeleur, lost his fortune in a gambling transaction in Dublin, and fled in disgrace unable to pay his debts. The persons who took over the estate under bankruptcy proceedings refused to recognise the community, insisted upon treating its members as common labourers on the estate, seized upon the buildings and grounds and broke up the Association.

So Ralahine ended. But in the rejuvenated Ireland of the future the achievement of those simple peasants will be dwelt upon with admiration as a great and important landmark in the march of the human race towards its complete social emancipation. Ralahine was an Irish point of interrogation erected amidst the wilderness of capitalist thought and feudal practice, challenging both in vain for an answer. Other smaller communities were also established in Ireland during the same period. A Lord Wallscourt established a somewhat similar community on his estate in County Galway; *The Quarterly Review* of November, 1819, states that there was then a small community existent nine miles outside Dublin, which held thirty acres, supported a priest and a school of 300 children, had erected buildings, made and sold jaunting cars, and comprised butchers, carpenters and wheelrights; the Quakers of Dublin established a Co-operative Woollen Factory, which flourished until it was destroyed by litigation set on foot by dissatisfied members who had been won over to the side of rival capitalists, and a communal home was established and long maintained in Dublin by members of the same religious sect, but without any other motive than that of helping forward the march of social amelioration. We understand that the extensive store of Messrs. Ganly & Sons on Usher's Quay in Dublin was the home of this community, who lived, worked and enjoyed themselves in the spacious halls, and slept in the smaller rooms of what is now the property of a capitalist auctioneer.

Appendix II

THE OWENITE COMMUNITY AT RALAHINE, COUNTY CLARE, 1831-33: A REASSESSMENT[1]

CORMAC Ó GRÁDA

From a late twentieth-century vantage point, the notion that pre-Famine Ireland, with its backward and unstable economic structure, was no place for piecemeal social experiments, may seem quite reasonable. But, as is well known, both legislators and classical-minded economists took a rather passive, pessimistic attitude to a more drastic restructuring of the system. In general, little was done and the ruling ideology of the day made it difficult to do anything. So it is hardly surprisng that there were well-meaning individuals in Ireland who attempted local intervention and experimentation. Two in particular stand out because they professed – and tried out – ideas which made them quite famous throughout the land. These are Jeremy Bentham's most prominent left-wing follower, William Thompson of Rosscarbery, and the Owenite landlord from County Clare, John Scott Vandeleur. Thompson's career has been discussed in several different places, most exhaustively in Pankhurst's monograph. He must have been an exceptional individual, at once respected, if not revered, by the London circle of utilitarian philosophers, and idolized by the labourers and smallholders of Glandore and Rosscarbery. However, while his works continue to be studied and reissued, his influence on later social thinking, particularly that of Karl Marx, has been somewhat exaggerated.[2] Thompson's friend, Vandeleur, was in great vogue early in the present century, when the rural co-operative movement was at its most dynamic. James Connolly, for instance, devoted an entire chapter of *Labour in Irish History* to the story of Vandeleur's agrarian 'commune', and Professor O'Brien also discussed it at some length in his *Economic History of Ireland*. Both heartily approved in their way: to Connolly it was a 'partial experiment in Socialism', to O'Brien, 'an unqualified success . . . (which) would have continued to succeed, had it not been broken up by causes completely outside the control of the members'.[3] Since then less has been heard about Vandeleur, but the comprehensive account in R.G. Garnett's recently published *Co-operation and the Owenite Socialist Communities* makes further detailed narratives of the experiment at Ralahine superfluous.[4] The story of how this Clare proprietor introduced a particular brand of communalism on part of his estate, and then lost all he possessed in a gambling bout in Dublin two years later, has by now been retold sufficiently often. As its title implies, this latest account discusses Vandeleur's 'new system' in the context of Owenite experimentation elsewhere; evidence is adduced for the same combination of paternalism, enlightened self-interest, and idealism, all essential elements of Robert Owen's schema. Under the agreement of association, Vandeleur was chairman of the committee of nine that administered the small community – fifty-two members were admitted in all – and the stock, equipment and land were held in common from him, on the understanding that gradually these would be purchased on a communal basis. Not only does Garnett provide a fine narrative; he also presents a definitive bibliography of previous versions. This article will not attempt yet another chronological account, though part of the story is told *en passant*. It will deal rather with a number of interpretative points arising mainly out of Garnett's description. It differs from other accounts in its emphasis on the economic basis of the agreement rather than on its political inspiration.

I

One deep-rooted misconception regarding the pre-Famine economy, which recent Irish historical scholarship has been correcting, is the notion that the countryside was populated almost exclusively by some ten thousand landlords and literally millions of tenants on miniscule holdings. Given that the Poor Inquiry and the Devon Report – the primary sources most often quoted – were concerned mainly with the problems of the labourers and cottiers, the error is easier to understand. In reality, however, there was a substantial farming community in between, largely overlooked by both contemporary commentators and historians. There were places too, especially in the west, where semi-proletarian status and small holdings predominated. Vandeleur's own Ralahine, a demesne townland on the Limerick-Ennis road, seems to have been one such place. The relations of production there before the experiment were probably typical of areas where medium-size farms were relatively rare: labourers would work off the high rent of the conacre for a proprietor or large farmer dealing in labour-intensive produce. When tillage prices fell or when the potato crop failed, the position of the conacre-man was desperate; in such situations rural revolt and food riots reached a peak.

Why more trouble might have been expected in County Clare than elsewhere may be guessed from the figures presented in Table I. These, taken from a pamphlet of 1831 or

TABLE I

Land and Wealth Distribution in Pre-Famine Clare

	Classes (by value of land)	*No. of Families in each class*	*No. of persons, house servants included*	*Average quantity of land possessed by each family (Plantation acres)*	*Total quantity of land possessed by each class*
I	£100	1,000	12,000	120	120,000
II	£20–100	3,000	21,000	30	90,000
III	£10–20	10,000	50,000	8	80,000
IV	Class possessing a little land	16,000	80,000	4	64,000
V	No possessions	16,000	80,000	waste lands	122,000
		46,000	243,000		476,000

1832, seem reliable; and, by the way, they corroborate recent criticisms made of the agricultural statistics given in the Irish census report of 1841.[5] Bearing in mind that an English acre is slightly over six-tenths of the traditional Irish measure, it would seem that well over three quarters of the holdings in pre-Famine Clare were less than fifteen statute acres, while the national proportion was less than two-thirds. Vandeleur must not have been alone there in fearing hungry, rebellious labourers. Indeed, the assassination of the land steward at Ralahine seems to have provoked the proprietor into introducing the 'new system' earlier than he had intended: Owenism, long a fancy of Vandeleur's, in the end may have been brought in for largely practical reasons.

Ralahine was not Vandeleur's only property in County Clare; he also owned another nearby which was leased out to tenant farmers. For whatever reason, it played no part in this exercise in Owenite experimentation. It was surely no accident that Vandeleur set about introducing his ideas in a setting of demesne agriculture rather than one of leasehold farming. Vandeleur's other property was larger than Ralahine, but it never receives a mention in the context of the 'new system'. The reason was probably that the introduction of the 'system' on leased land would have been both difficult and unnecessary. Vandeleur's whole problem was with labourers rather than with tenants. In modern idiom, it was a labour-management problem. The temporary success of Ralahine can hardly be regarded, therefore, as a 'triumph for fixity of tenure', as the late Professor George O'Brien would have it.[6] After all, the complaints of the cottier and labourer of the early 1830's were quite different from the demands of the Land League tenants forty or fifty years later. Besides, though the early ideologists of the Irish co-operative movement used the story to some effect, the context was also entirely different. AE knew this: 'conditions have changed in Ireland', he said. 'It would be impossible to use the rules devised by Craig in a co-operative community today'.[7] The reminder that both Vandeleur's troubles and the solution he proposed reflected the socio-economic conditions of the day is perhaps unnecessary. 'Utopian socialist' schemes are inevitably (and paradoxically) constrained in this way. This is not to deny the Ralahine version its purely Owenite trappings: the use of labour notes and the emphasis on education, discussed in almost all accounts of Ralahine, are taken straight from Owen's own blue-prints.

Edward Thomas Craig, Vandeleur's faithful lieutenant and organiser at Ralahine, called the story 'a romance of facts and figures'. Fortunately for us, Craig himself left behind in print several useful tables of figures. These are published in his own reliable, if highly eccentric, *History of Ralahine*, which he wrote when an old man.[8] The figures are informative in several respects. For one thing, they show how remarkably commercial was the character of agriculture at Ralahine. An examination of the crop acreages given by Craig and of Vandeleur's annual rental account shows this quite clearly. Taking the wheat crop first, Tables II and III show that the acreage under wheat was over sixty-five statute acres, while Vandeleur's claim was 320 barrels a year. If P.M.A. Bourke's pre-Famine yield data are accepted, this quanitity of wheat would have required sixty-four acres

TABLE II

Produce to be raised for rent

Item	*Price*	*Value*
	per stone	
6,400 stones of wheat at	1s 6d	£480
3,840 ,, ,, barley at	10d	£160
480 ,, ,, oats at	10d	£20
	per cwt	
70 cwt ,, beef at	40s 0d	£140
30 ,, ,, pork at	40s 0d	£60
10 ,, ,, butter at	80s 0d	£40
	Total:	£900

Source: Craig (1920), p.95.

TABLE III

Land Allocation at Ralahine (English Acres)

	A	R	P
Wheat	65	2	0
Barley	24	1	0
Oats	32	3	8
Potatoes	38	3	0
Turnips	30	3	4
Mangel-Wurzel	6	1	36
Vetches	15	3	6
Clover & Rye Grass	30	0	24
Fallow	23	1	27
Pastures & Plantation	280	0	7
Bog	63	2	26
Orchard	3	3	30
Houses	2	1	28
Total:	618	0	26

Source: Craig (1920) pp. 85–6.

of average land at the time.[9] Similarly, the barley acreage at Ralahine was about twenty-four, and Vandeleur's annual share 480 hundred-weight. Bourke's yield information implies an acreage of twenty-eight acres. Assuming higher than average yields, the allotted fields at Ralahine must still have been hard put to produce the required quantities of grain. It seems only fair to conclude, then, that all the barley and nearly all the wheat at Ralahine was marketed. This accords well with the qualitative evidence. But barley and wheat alone accounted for almost half the annual value added at Ralahine, and for over two-thirds of the rent payable. The tenants also undertook to supply Vandeleur with definite quantities of oats, butter, pork, and beef, all of which was sold in Limerick or Britain.[10] The oats crop at Ralahine was only partly marketed; perhaps as little as a sixth of the total. But the eleven horses – assuming they required, say, a stone a day each for half the year, and grazing for the other half – would have accounted for the bulk of the crop, and the rest would have gone on seed, hens, and oatmeal. The potato crop, also very large, was entirely unmarketed.

Detailed breakdowns like that provided by Craig are rare for the pre-Famine period, and there is the understandable temptation to over-generalize from them. However, assuming crop proportions at Ralahine did not change much with the 'new system' – and there is no strong evidence that they did – the size of the marketed share makes it difficult to support the dual economy argument that posits rigidly separated subsistence and commodity farming sectors, at least in this instance. Ralahine was backward, rural, and most of the labourers were monoglot Irish speakers, but they still worked on land that was producing for the market. In this Ralahine was probably not very different from the rest of Clare.[11]

Secondly, Vandeleur demanded rent worth £900 at 1830/1 prices out of an estimated total annual output of £1,700. As explained by Garnett, the £900 was a mere accounting figure, since the laboureres' obligations were to be met in kind (see Table II). Actually, Vandeleur's commodity bundle was probably worth less in 1832 and 1833 than it would have been in 1830 or 1831, grain prices having dropped in the meantime,[12] but his share of the total was still greater than half. Craig tells us that 'one or two of the members thought the bargain somewhat too favourable for the landlord,' and well they might; the landlord, too, admitted that the rent was 'too high'.[13] 'It must not be supposed', wrote Connolly, rather typically, 'that the landlord . . . had allowed his enthusiasm for Socialism to run away with his self-interest.'[14] I have calculated elsewhere that total rent payments for the whole of pre-Famine Ireland accounted for about £12 million out of an aggregate agricultural output of some £45 million; so it turns out that the proportion going to the proprietor at Ralahine substantially exceeded the national average.[15] Not only did the labourers undertake to supply the quantities specified; they also bore the cost of delivering the cereals to Limerick and the livestock to Liverpool. Craig tells of sixteen stall-fed cattle being brought all the way to Dublin 'at a considerable loss of fat' by two Ralahine drovers. The distribution costs involved here remain problematical; to take one instance, the labour-sheet for the work-week ending January 14, 1832, included ten shillings and eightpence for the expense of conveying wheat to Limerick.[16]

Taking all the above into account, the charge that Vandeleur may have milked the labourers for what he could seems plausible, if not proven. AE's 'flighty ideals'[17] had a less appealing, hard-headed side to him as well. And there is a further possible reason for suspecting Vandeleur: why should the Terry Alts, the rural labourers' secret army, want to kill his steward in the first place? The Terry Alts were what E.P. Thompson would call 'an army of redressers',[18] and generally selected their victim carefully. Surely this means that Vandeleur or his steward had antagonized and over-exploited the Ralahine workers before the experiment, in spite of the former's Owenite ideas?[19] Vandeleur's early interest in reform seems to have been of no avail – more a dilettante's interest perhaps?

II

A reading of all the data available on Ralahine suggests that neither rates of pay nor consumption patterns there differed much from those of similar areas in the west at the time. The daily pay was eightpence for a man and fivepence for a woman, and the hours were regulated by the seasons. A few skilled men were paid extra. Such rates were common in normal times in pre-Famine Ireland. The labourers' food continued to be largely potatoes and milk. The potato acreage at Ralahine, just less than forty statute, allowed a per capita consumption of the order usual for the poor in Ireland as a whole. Here again the detail on yields is lacking. If one assumes a level of six tons per acre, then a consumption level of twelve pounds per adult male per day, with comparable allowances for the women and the young, would have left well over one hundred tons for non-human consumption.[20] The Ralahine pigs presumably consumed most of this. Milk consumption at Ralahine was very high, certainly higher than the national average. However, nothing is known about milk consumption there before the 'new system'. True, some meat was sold to members: the records show a weekly sale of 9½lbs of mutton on one bill, and of sixty pounds of pork on another. But from Craig's own qualitative remarks, it is safe to assume that meat continued to be a luxury: 'our members lived chiefly on vegetables, consisting mainly of potatoes, which could not sustain the muscular vigour of the men without the milk . . . but scarcely any animal food'. Bread was non-existent in the everyday diet; at the harvest festival of 1832 'bread made under the directions of Mrs Craig (was) liberally supplied . . . . To the members the art of making their own bread was a novelty and a mystery'.[21]

What was so special then about Ralahine? Firstly, in spite of the high rent, the long hours, and the uninteresting diet, the 'new system' conferred one crucial benefit – work was plentiful and year-round. And while Vandeleur's terms were far from generous, they made the work less alienating than straightforward wage-labour. There was both the pressure to make the rent through co-operative effort, and the realization that any surplus above it would be the labourers' own. As Craig put it, 'the industry and skill of hitherto apathetic, sluggish, and sullenly silent labourers became manifest, affording a gratifying testimony to the beneficial influence of the plan which awakened the spirit of the workman, by giving him an impelling motive for exertion, by assuring him a share – although a small one – in the profits arising out of his increased exertions'.[22] In this context, a famous Ralahine incident bears further retelling:[23]

> During the winter of 1832, a hunted fox crossed the mill-course near the rickyard and took across the orchard

> and over a seventy-acre field of young wheat, in the best state of tillage. The field of mounted huntsmen were seen galloping towards the farmyard gates, which were solid timber, some eight feet high, fixed on rollers running on rails. By a sudden impulse the members in the farmyard closed the gates, and brought the huntsmen abruptly to bay, and thereby put a stop to the chase for the day. The horsemen were astonished at the presumption of the men of the 'new system', who had thus dared to interfere with their sport. The men who composed this wild group, chasing a poor frightened animal, were farmers, tradesmen from the neighbouring towns, clergymen and 'squireens'. The latter were the younger sons of respectable or wealthy families, like the brother of the proprietor at Ralahine, having little or no patrimony of their own, but who scorned to demean themselves by any profitable occupation. Many of the huntsmen loudly expressed their disappointment and indignation at the rudeness of the labourers daring to talk to them about 'young crops' and 'damaged fences'! I had repeatedly heard some of the younger members express themselves strongly against the practice of hunting over tillage land. I had given no advice on the matter. Had Mr Vandeleur been on the spot he would, doubtless, have permitted the horsemen their gallop over the wheatfield after the hounds and poor Reynard, who, on that occasion carried his brush home in safety.

Finch, who scrutinized the 'system' at work, tells us that there were a few idlers about the place at first, but that the 'committee knew their characters, and appointed one of these idlers to work between two others who were industrious — at digging, for instance: he was obliged to keep up with them, or he became the subject of laughter and ridicule to the whole society.'[24] In a limited way, then, Ralahine showed that a less hierarchical work structure *per se* did not necessarily affect the level of output. The system also meant that some agricultural improvement and building would proceed; subject to the rent contraint, the labourers, through an elected committee, could choose how to allocate their work-time on the demesne. From the evidence, the landlord invested nothing out of his annual rent on improvement in 1831-3, while the labourers spent several hundred pounds of their share on repairs, seed, and new investment. Actually, the commitment to full employment for all may have been responsible for the investment, since there was little scope for more labour-intensive cultivation or direct output-augmentation in the short run. Thus improvement had less to do with security of tenure – which was not part of the agreement in any case – than with full employment.

III

Finally, there is AE's verdict that 'when John Scott Vandeleur gambled at his club he gambled away not merely his own property, but what well may have been a happier destiny for his country'.[25] How should one treat this rather dramatic counterfactual statement? One can think of two plausible answers, the first rather sceptical, the second more optimistic. Let us deal first with the negative side. To begin with, despite the extent of subsequent documentation, Ralahine attracted little notice at the time of the experiment. There were, predictably, visits by Robert Owen, William Thompson, and others, and one or two landlords showed some interest, but the experiment was completely overlooked in the local and national press of the day. Needless to say, there were no permanent spread effects whatever. Perhaps contemporaries were inclined to regard the 'new system' as yet another 'farming improvement', and were blind to any wider social or political implications? Perhaps no one was interested enough to report on, far less to replicate, a largely untried 'new system' elsewhere?

Secondly, one can make the case that Ralahine itself

would not have lasted long in the form set out by Vandeleur. 'Might-have-been' guesses are cheap, but the following may not seem too far fetched. The occasion for introducing the 'new system', as already pointed out, was rural revolt and widespread hunger in the immediate County Clare region. Thus Vandeleur's rental income was relatively low in 1829-31 because of the miserable weather and the resultant bad harvest. 'The years from 1829 to 1831', we are told, 'were "famine years" in Ireland, complicated by the severe economic distress already prevailing in both England and Ireland. By August 1831, one observer could describe the condition of the peasants as "more wretched than I could possibly imagine had I not witnessed it." '[26] Now the harvests of 1832 and 1833, the period of the experiment, were extremely good by comparison, and quite capable of yielding the surplus demanded by the landlord. There is no direct quantitative evidence on the bounty of the cereal crops at Ralahine in 1832-3, but the spirit of the harvest celebrations, and an indirect remark about the orchard, would suggest excellent results.[27] The members were not to experience the hardships of a bad harvest for as long as the 'system' lasted. However, 'new system' or no 'new system', it is doubtful whether Ralahine could have provided Vandeleur with his commodity bundle in another bad stretch like 1829-31. As shown earlier, Vandeleur's demands implied average, or better than average, yields for corn crops. A bad summer, which could easily result in a twenty or thirty per cent decline in yields, would have made such claims absolutely impossible to meet. Nor is it likely that the labour force at Ralahine could have cushioned themselves against bad harvests rapidly enough, by cultivating the required extra acreage annually: they were sufficiently hard pressed at seasonal peaks as it was.

On the basis of Table IV, a plausible index of pre-Famine swings in Irish agricultural output, Ralahine might well have lasted through most of the thirties. However, the same table implies that, *given the proprietor's terms*, the experiment

TABLE IV

Grain and Malt Exported from Irish ports to Great Britain, 1828 – 1841

Year	1,000 *Quarters*	*Year*	1,000 *Quarters*
1828	2826	1835	2679
1829	2306	1836	2908
1830	2216	1837	3030
1831	2429	1838	3474
1832	2991	1839	2243
1833	2737	1840	2328
1834	2763	1841	2856

Source: A Statement of the Quantities of each kind of Grain and of Malt imported into Great Britain from Ireland, in each year from 1800 to 1841, both inclusive (Brit. Parl. Papers, 1842, XL).

would have had difficulty in surviving the crisis of 1839-40. If this is so, AE's claim seems both misleading and exaggerated. Yet one might also maintain, in support of AE, that arrangements such as Ralahine, given time, create their own inward dynamic. Either sufficient pressure from the labourers, or resourcefulness on Vandeleur's part, might have carried the 'system' through such a crisis. The fragility of the 'system' was clearly shown in 1833, and the evidence seems against long-run stability. Nevertheless, it would be rash to reject entirely AE's counterfactual vision.

FOOTNOTES

1 I wish to thank Pól Ó Duibhir, Moore McDowell and Pádraig Ó Snodaigh, as well as several students at U.C.D. for their help and comments. I am particularly grateful to Dr Ronald Garnett for putting me right on several points. The conventional disclaimer naturally applies.
2 R.K.P. Pankhurst *William Thompson, 1755 - 1833: Britain's Pioneer Socialist, Feminist and Co-operator* (London, 1954); there is an in-

teresting, if brief account by Patrick Lynch in J.F. Boyle (ed.), *Leaders and Workers* (Cork, 1963), pp. 9-16. I am basing my view of Thompson's philosophy on a reading of *The Distribution of Wealth* (1825) and *Labour Rewarded* (1827). The latter is an attack on Hodgskin, who was closer in outlook to Marx than either Owenites or Benthamites.

3 J. Connolly, *Labour in Irish History* (Dublin, 1967), pp. 79-93; G. O'Brien, *The Economic History of Ireland from the Union to the Famine* (London, 1921), pp. 115-121.

4 Manchester, 1972, pp. 100-129.

5 'Agricola', *Considerations addressed to the Landed Proprietors of the County of Clare* (Limerick, 1832 [?]), p. 44. Quite coincidentally, this pamphlet is addressed to a certain James Molony of Kiltanon, Mrs Vandeleur's brother. After Vandeleur's disappearance, Molony became guardian at Ralahine.

Both the population statistics and the aggregate acreage given are quite plausible. Indeed, the population estimate is probably closer to the truth than the 258,000 reported in the official census of 1831. The farm size breakdown, adjusted to statute acres, agrees quite well with P.M.A. Bourke's amended figures in 'The Agricultural Statistics of the 1841 Census of Ireland: A Critical Review', *Economic History Review*, 2nd ser. xviii (1965).

6 O'Brien, op. cit. p. 120. Prof. O'Brien's reference to Craig as 'one of Owen's collaborators in Scotland' is inaccurate, as is the claim (p. 118) that 'wheaten bread took the place of potatoes as the staple food of the people'. This is explained below.

On the full employment aspect of Ralahine, see W. Pare, *Co-operative Agriculture* (London, 1870), pp. xviii, 53-4. Ronald Garnett has informed me that after 1823 Vandeleur set about removing small tenants at Ralahine to holdings 'more distant from his residence until his own farm increased to 700 acres.'

7 Foreword by AE (George Russell) to the 1920 edition of E.T. Craig, *An Irish Commune* (Dublin, 1920), p.v.

8 E.T. Craig, *The Irish Land and Labour Question, Illustrated in the History of Ralahine* (London, 1893). This work was reissued in abridged form in Dublin in 1920 (see above).

9 P.M.A. Bourke 'The Average Yields of Food Crops in Ireland on the Eve of the Great Famine', *Journal of the Department of Agriculture*, lxvi, 4.

10 For a contemporary account of Limerick and its port see H.D. Inglis, *Ireland in 1834* (London, 1834), I, pp. 296-7.

11 The dual economy hypothesis is forcefully advanced in J. Lynch and P. Vaizey, *Guinness's Brewery in the Irish Economy* (Cambridge, 1960), pp. 9-36, 161-176.

12 B. Mitchell and P. Deane, *Abstract of British Historical Statistics* (Cambridge, 1962), p. 488. While the agreement specified the quantities mentioned, it also read that 'in case Mr Vandeleur wishes, he may take labour notes in lieu of the above articles, which notes are to be received in the store for articles, at the same rate as they are sold to any of the members'. But William Pare suggests that 'if the quantity of any particular kind of produce should fall off, its equivalent might be given to the landlord in some other kind of goods . . . The fact was, however, that . . . the rent was paid in the six articles named'. See William Pare, *Co-operative Agriculture* (London, 1870).
13 Craig, (1893), p. 55.
14 Connolly, op. cit. p. 90.
15 Craig (1920), pp. 61, 95. The aggregate figures are discussed in C. Ó Gráda, 'Pre-Famine and Post-Famine Agricultural Head Rents', *Economic and Social Review*, (1974), 385-92.
16 Craig (1920), p. 61.
17 AE in Craig (1920), p. iv.
18 E.P. Thompson, *The Making of the English Working Class* (London, 1963), part 3.
19 Accordingly to Pare (op. cit. p. 13) Hastings, Vandeleur's steward, did treat the labourers cruelly. Sadistic acts, such as depriving them of water during heavy harvest work, 'an excuse for the reapers wasting their time', apparently drove the labourers into plotting his assassination. Why a supposedly humanitarian landlord stood idly by is difficult to explain.
It seems that Hastings had vigorously supported Vandeleur's efforts at introducing machinery in 1830/1, but was against any scheme along Owenite lines. Since the Terry Alts opposed the early moves towards mechanisation rather than the purely 'co-operative' side of Ralahine, the steward's assassination may well have been partly for 'luddite' reasons. The steward was killed in April 1831; the agreement was signed in November 1831. See Garnett, op. cit. pp. 105-6; Pare, pp. 15-32; letters from R.G. Garnett to author.
20 The figures used here are based on P.M.A. Bourke, 'The Use of the Potato Crop in Pre-Famine Ireland', *Journal of the Social and Statistical Inquiry Society*, xxi (1967-8), pp. 72-96. Bourke estimates that the average small peasant family consumed over four tons annually. Pigs consumed on average twelve pounds daily for 10½ months of the year.
21 Craig, (1893), p. 167.
22 Craig, (1920), p. 134.
23 Craig, (1893), 138-9. Craig's work follows very closely indeed Pare's *Co-operative Agriculture*. It is not clear, however, whether Craig is based on Pare, or *vice-versa*. Pare himself tells us (op. cit. p.

xxiii) that 'by way of making my materials more full and accurate, I have, much to his gratification, put Mr Craig under a written cross-examination, so to speak, as to all the more important details in the course of which it has been pleasing to observe the great care and scrupulosity with which his answers have been given'. Compare Craig's account of the fox-hunt with the following from Pare:

'During the winter of 1832, a hunted fox crossed the mill water course near the rick-yard, and took across the orchard . . . The mounted huntsmen – young squires, farmers, and tradesmen – to keep well up with the hounds on the wheat field, would have to pass through the farm-yard; but they found that, by a sudden and mutual impulse, the large high gates of the farm-yard had been locked against them by the "New Systemites" . . . Many of the huntsmen seemed perfectly astounded at the daring and "impudence" of the "New System" men. Had Mr Vandeleur been at home, he would doubtless have allowed the huntsmen their gallop over the wheat field.'

Perhaps Craig felt entitled, in old age, to re-use liberally the same information that he had previously passed on to Pare? A charge of straight plagiarism would be unfair, then.

24 Pare, op. cit. p. 61.

25 AE, loc. cit. p. iv.

26 G. Broeker, *Rural Disorder and Police Reform in Ireland, 1812-1836* (London, 1970), p. 204.

27 See Craig, (1920), p. 86; Craig, (1893), pp. 165-7; Pare, op. cit. pp. 112-14.

Some other publications in print related to Co-operation in Ireland

---

HOW TO START A CO-OP
Mary Linehan, Con O'Leary, Vincent Tucker (1981), £0.50p

CO-OPERATIVES AND THE LAW
Edited by Mary Linehan (1982), £2.00

CO-OPERATIVE HOUSING IN IRELAND
Edited by Vincent Tucker (1982), £2.00

CO-OPERATION BETWEEN CO-OPERATIVES
Michael Ward, Robert Briscoe and Mary Linehan (1982), £3.50

THE CO-OPERATIVE IDEA
Briscoe et al. (1982)

WORKERS CO-OPERATIVES: POTENTIAL PROBLEMS
Edited by Mary Linehan and Vincent Tucker (forthcoming) (1983)

Postage extra.

Available from:
Bank of Ireland Centre for Co-operative Studies,
University College, Cork.

THE IRISH CO-OPERATIVE MOVEMENT:
ITS HISTORY AND DEVELOPMENT
Patrick Bolger (1977)

Available from:
The Institute of Public Administration,
59 Lansdowne Road, Dublin 4.
paperback £4.50, hardback £7.50

A STUDY OF WORKER'S CO-OPERATIVES
Robert O'Connor and Phillip Kelly (1980)

Available from:
The Economic and Social Research Institute,
4 Burlington Road, Dublin 4.
£5.00 (special rate of £2.50 for bona fide students).

THE FIRST EIGHTY YEARS: A History of Monaghan Co-op
(1983), £5.00

FRAMEWORK FOR CO-OPERATIVE DEVELOPMENT
(1979), £0.75

Available from:
Irish Co-operative Organisation Society Ltd.,
The Plunkett House, 84 Merrion Square, Dublin 2.
